Statistical Thermodynamics

An Engineering Approach

Dr. Daily is currently Professor of Mechanical Engineering at the University of Colorado at Boulder. He studied mechanical engineering at the University of Michigan (BS 1968, MS 1969) and at Stanford University (PhD 1975). Prior to starting college he worked on sports and racing cars, owning his own business. Between the MS and PhD degrees he worked as a heat transfer analyst at Aerojet Liquid Rocket Company. After receiving the PhD he was a faculty member at the University of California at Berkeley until 1988, when he moved to the University of Colorado. He has served as the Director of the Center for Combustion Research and as Chair of the Mechanical Engineering Department at the University of Colorado.

His academic career has been devoted to the field of energy, focusing on combustion and environmental studies. He has worked on combustion and heat transfer aspects of propulsion and power generation devices, studying such topics as fluid mechanics of mixing, chemical kinetics, combustion stability, and air pollution. He also works on the development of advanced diagnostic instrumentation (including laser based) for studying reacting flows and environmental monitoring. Most recently he has been working in the areas of biomass thermochemical processing and source characterization, wildfire behavior, the environmental consequences of combustion, and optical biopsy of cancer. He is a founder of Precision Biopsy Inc., a company developing technology for the optical detection of prostate cancer.

Dr. Daily served as a member of the San Francisco Bay Area Air Quality Management District Advisory Council for 10 years. He served on and chaired the State of Colorado Hazardous Waste Commission for over 10 years and was on the State of Colorado Air Quality Control Commission. He is a Fellow of The American Institute of Aeronautics and Astronautics (AIAA) and serves as chair of its Publications Committee.

Statistical Thermodynamics

An Engineering Approach

JOHN W. DAILY
University of Colorado Boulder

CAMBRIDGE
UNIVERSITY PRESS

CAMBRIDGE
UNIVERSITY PRESS

University Printing House, Cambridge CB2 8BS, United Kingdom

One Liberty Plaza, 20th Floor, New York, NY 10006, USA

477 Williamstown Road, Port Melbourne, VIC 3207, Australia

314–321, 3rd Floor, Plot 3, Splendor Forum, Jasola District Centre, New Delhi – 110025, India

79 Anson Road, #06–04/06, Singapore 079906

Cambridge University Press is part of the University of Cambridge.

It furthers the University's mission by disseminating knowledge in the pursuit of education, learning, and research at the highest international levels of excellence.

www.cambridge.org
Information on this title: www.cambridge.org/9781108415316
DOI: 10.1017/9781108233194

First published 2019

Printed and bound in Great Britain by Clays Ltd, Elcograf S.p.A.

A catalogue record for this publication is available from the British Library.

Library of Congress Cataloging-in-Publication Data
Names: Daily, John W. (John Wallace), author.
Title: Statistical thermodynamics : an engineering approach / John W. Daily (University of Colorado Boulder).
Other titles: Thermodynamics
Description: Cambridge ; New York, NY : Cambridge University Press, 2019. | Includes bibliographical references and index.
Identifiers: LCCN 2018034166 | ISBN 9781108415316 (hardback : alk. paper)
Subjects: LCSH: Thermodynamics–Textbooks.
Classification: LCC TJ265 .D2945 2019 | DDC 621.402/1–dc23
LC record available at https://lccn.loc.gov/2018034166

ISBN 978-1-108-41531-6 Hardback

Brief Contents

Contents

List of Figures

List of Tables

Preface

I have been teaching advanced thermodynamics for over 40 years, first at the University of California at Berkeley from 1975 through 1988, and since then at the University of Colorado at Boulder. I have mostly had mechanical and aerospace engineering students who are moving toward a career in the thermal sciences, but also a goodly number of students from other engineering and scientific disciplines. While working on my Master's degree at the University of Michigan I took statistical thermodynamics from Professor Richard Sonntag using his and Gordon Van Wylen's text *Fundamentals of Statistical Thermodynamics* [1]. Later at Stanford I took Charles Kruger's course using *Introduction to Physical Gas Dynamics* by Vincenti and Kruger [2]. This course had a large dose of statistical thermodynamics combined with gas dynamics.

Both experiences sharpened my interest in the subject matter. Then, early in my teaching career, I had the good fortune to be introduced to the wonderful text *Thermodynamics and an Introduction to Thermostatistics* by Professor Herbert B. Callen [3] of the University of Pennsylvania. My first reading of his postulatory approach to classical thermodynamics was one of the most exciting learning experiences of my career. As one epiphany followed another, I realized that here was a teacher's dream, the opportunity to teach classical thermodynamics in a way that makes Gibbs' [4] ensemble approach to statistical thermodynamics transparent. I have therefore taught advanced thermodynamics in this fashion over the years, but with a serious handicap. There was no available text that was suitable for engineering students and that fully integrates Callen's postulatory approach with the ensemble approach of statistical mechanics. Eldon Knuth's book *Introduction to Statistical Thermodynamics* [5] is a good companion to Callen, but uses different notation. Tien and Lienhard's *Statistical Thermodynamics* [6] introduces the postulatory approach, but in a very short first chapter. They also spend a great deal of time on classical statistics, which I feel is unnecessary when ensemble statistics are used. All are quite dated, especially in terms of advances in computation techniques and modern property compilations. I also feel indebted to Frank Incropera's book *Introduction to Molecular Structure and Thermodynamics* [7], which provides a particularly easy to follow presentation of quantum mechanics suitable for engineering students.

Hence this book. It assumes the reader is a mechanical or aerospace engineering graduate student who already has a strong background in undergraduate engineering thermodynamics and is ready to tackle the underlying fundamentals of the subject. It is designed for those entering advanced fields such as combustion, high-temperature gas dynamics, environmental sciences, or materials processing, or who wish to build a background for understanding advanced experimental diagnostic techniques in these or similar fields. The presentation of the subject is quite different from that encountered in engineering thermodynamics courses, where little fundamental explanation is given and the student is required to accept concepts such as entropy and the 2nd Law. Here, the underlying meaning of entropy, temperature, and other thermodynamic concepts will be definitively explored, quantum mechanics learned, and the physical basis of gas, liquid, and solid phases established. In addition, the molecular basis of transport phenomena and chemical kinetics will be explored. Modern tools for solving thermodynamic problems will also be explored, and the student is assured that he or she will gain knowledge of practical usefulness.

Comment on Software

In a number of locations throughout the text, various software programs will be mentioned. Some are open source, others commercial. Two packages are mentioned multiple times: Mathcad and Mathematica. Both are commercial but almost all universities have site licenses for engineering students and student licenses are very affordable. At the University of Colorado we have favored Matlab for many years, and it is expected that students will be adept in its usage. Where other commercial programs are mentioned, there is almost always an open source alternative given as well. As is usually the case, the commercial programs are more polished, with easier to use interfaces. However, the open source programs can work well and in some cases the science is more up to date. I realize that in this day and age electronic anything tends to come and go. I have tried to reference software that is likely to have staying power for some time. However, it is incumbent on any engineer or scientist to stay current on available tools, so I expect that the conscientious student (and teacher) will find suitable alternatives if necessary.

Acknowledgments

In any endeavor such as writing a book of this nature, it is clear that one owes debts to a number of people. My start came from being blessed with being born into an academic family. My father, James W. Daily, studied at Stanford and Cal Tech, before teaching at MIT for 18 years, and later at the University of Michigan, serving as Chair of the Applied Mechanics Department. As a youth I met many giants in engineering and science, including G. I. Taylor, Hermann Schlichting, Theodore von Kármán, and Harold Eugene "Doc" Edgerton. I have already mentioned studying thermodynamics under Richard Sonntag at Michigan. One of my PhD advisors at Stanford was Charles Kruger. I also had classes from Bill Reynolds and Milton Van Dyke, both great minds. And while teaching at Berkeley I had many scintillating conversations with Rick Sherman, Chang Tien, George Pimentel, Yuan Lee, Bill Miller, and Henry "Fritz" Schaefer III. While at Boulder I have developed wonderful collaborations with G. Barney Ellison, John Stanton, Peter Hamlington, Melvin Branch, Greg Rieker, Nicole Labbe, and others. My many colleagues around the world have kept me honest through their reviews of papers and proposals, provided spirited discussions at meetings and meals, and generally challenged me to be my best. And needless to say, my graduate students have provided great joy as we transition from teacher–student into lifelong equals and friends.

At home my wife Carol has been an inspiration. As I write this she is battling ovarian cancer with courage and grace. I am not surprised. She raised four children while working as a teacher and psychotherapist helping children, all while managing to ski, backpack, run marathons, and compete in triathlons. We have been rewarded with ten wonderful grandchildren.

Of course, thanks go to the people at Cambridge University Press, including Steven Elliott and Brianda Reyes. One of our current graduate students, Jeff Glusman, was particularly helpful with proofreading and made many valuable suggestions.

To all these people I give my heartfelt thanks. Because of them I have had a great life that has given me the opportunity to write this book.

1 Introduction

1.1 The Role of Thermodynamics

With an important restriction, the discipline of thermodynamics extends classical dynamics to the study of systems for which internal, microscopic, modes of motion are important. In dynamics, we are concerned about the macroscopic motion of matter. In thermodynamics it is motion at the microscopic level that mostly absorbs our interest. In dynamics we use the concepts of kinetic and potential energy and work to describe motion. In thermodynamics we add to these concepts internal energy and heat transfer, along with related properties such as temperature, pressure, and chemical potential. The important restriction, which we will discuss in detail later, is that thermodynamics is limited to analyzing changes between equilibrium states. Systems for which internal modes of motion are important include power generation, propulsion, refrigeration, chemical processes, and biology.

The goal of applied thermodynamic analysis is to understand the relationship between the design parameters and the performance of such systems, including specification of all appropriate state properties and energy flows. This task may be cast in terms of the following steps:

1. Identify any process or series of processes including components of cyclic processes.
2. Select control masses or volumes as appropriate.
3. Identify interactions between subsystems (i.e. work, heat transfer, mass transfer).
4. Sketch a system diagram showing control surfaces and interactions and a process diagram showing state changes.
5. Obtain all necessary properties at each state given sufficient independent properties – for example u, v, h, s, T, p, and chemical composition.
6. Calculate interactions directly where possible.
7. Apply the 1st Law to any process or set of processes.
8. Calculate the behavior of an isentropic process or a non-isentropic process given the isentropic efficiency.
9. Put it all together and solve the resulting system of nonlinear algebraic equations.
10. Calculate the system performance, including 2nd Law performance.

Most of these tasks are addressed in undergraduate engineering thermodynamics, at least for systems involving simple compressible substances as the working fluid.

However, the level of conceptual understanding necessary to address more complex substances and to understand and carry out 2nd Law performance analysis is usually left for graduate study. This is the focus of this book.

1.2 The Nature of Matter

As we know, matter is composed of atoms and molecules. The typical small atom is composed of a positively charged nucleus and negatively charge electrons. The nucleus, in turn, is composed of positively charged protons and neutral neutrons. The charge on electrons and protons is 1.602×10^{-19} C. It is the electrostatic forces that arise from charged electrons and protons that hold atoms together, allow for the formation of molecules, and determine the overall phase of large collections of atoms and/or molecules as solid, liquid, gas, or plasma. The spatial extent of electrostatic forces for a typical small atom is approximately 5 Å or 5×10^{-10} m. There are about 2×10^9 atoms per lineal meter in a solid, resulting in about 8×10^{27} atoms/solidus m^3. Thus, macroscopic systems are composed of a very large number of atoms or molecules.

In macroscopic systems, we describe behavior using the equations of motion derived from Newton's Law. In principle, we should be able to solve the equations of motion for each atom or molecule to determine the effect of microscopic modes of motion. However, even if we ignore the fact that the behavior of individual atoms and molecules is described by quantum mechanics, it would be impossible to simultaneously solve the enormous number of equations involved. Clearly, an alternative approach is required as some kind of averaging must take place. Fortunately, nature has been kind in devising the laws of averaging in ways that allow for great simplification (although we will explore solving the classical equations of motion for small subsets of atoms as a way of estimating thermodynamic and other properties).

Thus, the solution of thermodynamics problems breaks down into two great tasks. The first is developing the rules for macroscopic behavior, given basic knowledge of microscopic behavior. We call this subject classical or macroscopic thermodynamics. Providing microscopic information is the subject of statistical or microscopic thermodynamics.

1.3 Energy, Work, Heat Transfer, and the 1st Law

The basis of the concepts of energy, kinetic and potential, and work can be derived from Newton's Law:

$$\vec{F} = m\vec{a} \tag{1.1}$$

Consider applying a force to a macroscopic body of mass m, causing it to follow some trajectory. Integrating Newton's Law over the trajectory, we obtain

$$\int_{\vec{x}_1}^{\vec{x}_2} \vec{F} d\vec{x} = \int_{\vec{x}_1}^{\vec{x}_2} m\vec{a} d\vec{x} = \int_{\vec{x}_1}^{\vec{x}_2} m\frac{d\vec{V}}{dt} d\vec{x} = \int_{\vec{V}_1}^{\vec{V}_2} m\vec{V} d\vec{V} = \frac{1}{2}m\left(V_2^2 - V_1^2\right) \tag{1.2}$$

We normally identify

$$W_{12} = \int_{\vec{x}_1}^{\vec{x}_2} \vec{F} d\vec{x} \tag{1.3}$$

as the work done during the process of causing the mass to move from point $\vec{x}_1$ to $\vec{x}_2$. The work will depend on the path function $\vec{F}(\vec{x})$. Indeed, different functions can result in the same amount of work. As a result, we say that work is a path, or process integral.

In contrast, the integral of $m\vec{a}$ depends only on the value of the velocity squared at the end points. We identify

$$KE = \frac{1}{2}mV^2 \tag{1.4}$$

as the kinetic energy. The energy is a point or state property, and the integral of $\frac{1}{2}mdV^2$ is an exact differential.

The concept of potential energy arises out of the behavior of a body subject to a potential force field. A potential field is one in which the force imposed on the body is a function of position only. Gravity is the most common example of a potential field encountered in practice. Consider the case where a body subjected to an applied force is in a gravitational field whose effect is the constant weight W. If the gravitational field operates in the z direction, then Newton's Law takes on the form

$$F_z - W_z = ma_z \tag{1.5}$$

In the absence of an applied force, $W_z = ma_z$. Defining g as the effective acceleration due to gravity, $W_z = mg$. Adding this to the applied force and integrating as above gives

$$W_{12} - mg(z_2 - z_1) = \frac{1}{2}m(V_2^2 - V_1^2) \tag{1.6}$$

Note that the integral of the potential force term is in an exact form, and depends only on the value of mgz at the end points. Therefore, we normally call

$$PE = mgz \tag{1.7}$$

the potential energy, and write Eqs (1.1)–(1.6) as

$$\frac{1}{2}m(V_2^2 - V_1^2) + mg(z_2 - z_1) = W_{12} \tag{1.8}$$

or

$$\Delta E = \Delta KE + \Delta PE = W_{12} \tag{1.9}$$

where

$$E = KE + PE \tag{1.10}$$

This is a statement of the 1st Law of Thermodynamics for a body with no internal energy. As can be seen, it means that energy can be viewed as a property that measures the ability of matter to do work. Furthermore, rather than the absolute value of the energy, the important quantity is the change in energy.

The concepts of work and energy can also be applied at the microscopic level. Atoms and molecules, and nuclei and electrons, can have kinetic energy, and the electrostatic fields within and between atoms can lead to potential energy. Furthermore, electrostatic forces can result in work being done on individual particles. If we identify the total kinetic and potential energy at the microscopic level as U, then the total energy of a macroscopic body becomes

$$E = U + KE + PE \tag{1.11}$$

Heat transfer is work carried out at the microscopic level. It arises from random individual molecular interactions occurring throughout a material or at a surface between two materials. The non-random, or coherent motion leads to macroscopic work, the random component leads to heat transfer. We typically identify heat transfer as Q, and the complete form of the 1st Law becomes

$$\Delta E = W_{12} + Q_{12} \tag{1.12}$$

or in differential form

$$dE = \delta W + \delta Q \tag{1.13}$$

where δ indicates that work and heat transfer are path functions, not exact differentials.

1.4 Equilibrium

As mentioned in the first paragraph of this book, thermodynamics involves the study of systems that undergo change between a set of very restricted states called equilibrium states. Equilibrium is the stationary limit reached by some transient process and one must establish that equilibrium is reached and thermodynamic analysis can be used. As we shall see, the statistics of large numbers lead to great simplifications when considering stationary or equilibrium states.

At the microscopic level processes are usually very fast. Typical internal motions such as rotation or vibration occur with periods of 10^{-12}–10^{-15} sec. In a gas, collisions occur every few 10^{-9}–10^{-12} sec. Thus, internal modes tend to want to be in equilibrium, at least locally. At the macroscopic level, however, processes such as flow, heat transfer, and mass transfer can be quite slow. Table 1.1 lists various processes and characteristic times for them to occur. When it is necessary to understand the system behavior while these transient processes are occurring, one must use the principles of fluid mechanics, heat transfer, mass transfer, and so on. (Here, L is a characteristic length, V a characteristic velocity, α the thermal diffusivity, and D the mass diffusivity.)

This leads to a very important concept, that of "local thermodynamic equilibrium" (LTE). If local molecular relaxation processes are very fast compared to

Table 1.1 Characteristic times of transport processes

Process	Characteristic time
Flow	L/V
Heat transfer	L^2/α
Mass transfer	L^2/D

the characteristic times of Table 1.1, then locally within the flow thermodynamic equilibrium will hold, allowing the use of all the normal thermodynamic relationships. Indeed, this approximation holds for almost all flow, heat transfer, and mass transfer processes that are normally encountered, with the exception of very highspeed flows.

1.5 Thermodynamic Properties

Properties are quantities that describe the state of a system. Position, rotational orientation, and velocity, for example, describe the instantaneous dynamic state of a solid body. In thermodynamics we are concerned with problems involving work and heat transfer. Therefore, energy must be an important property, as it is a measure of the ability of matter to do work or transfer heat. Work and heat transfer are dynamic quantities and not descriptive of the state of a system. However, the amount of work or heat transfer required to bring a system to a given state will clearly be influenced by the size of the system, and thus volume and mass are also important thermodynamic properties. Composition is also an important property, as it will clearly affect the microscopic behavior of a system. To summarize, we have identified energy, volume, and mass (or moles) as important properties. For a multicomponent system we must also specify either the mass or moles of each specie or phase present.

The properties thus identified, U, V, and N_i (where N_i is the number of moles of component i), have an important feature in common. They are all properties that are extensive in nature. By that we mean that they are a measure of the size of the system. And, in fact, were the size of a system to change, all other things being held constant, each of these properties would change by the same amount. We thus formally call them extensive properties. If, for a closed system, the extensive properties U, V, and N_i are specified, then the thermodynamic state is completely specified.

In addition to the extensive properties, we shall be concerned with several intensive properties. Intensive properties are properties that do not scale with the size of a system, but rather are a function of the normalized extensive properties. Temperature and pressure are examples of important intensive properties. As we shall see, the intensive properties are each paired with a related extensive property, and are defined in terms of the extensive properties.

The extensive properties can be cast in an intensive form by normalization, usually in terms of the volume or total mass or moles. However, as we shall see, they remain fundamentally different in character from the true intensive properties.

1.6 The Fundamental Problem of Thermodynamics

There are four types of thermodynamic problems. These are:

1. Initial state/final state problems. Involve specifying the initial states of two or more subsystems that may include reservoirs. Work, heat transfer, or mass transfer is then allowed and the final states of the subsystem determined.
2. Specified interaction problems. In specified interaction problems, one specifies the nature and value of interactions of a system with its surroundings. Consider compressing a gas in an insulated piston cylinder arrangement. If the initial state of the gas and the work done in compression are specified, then one can calculate the final state of the gas.
3. Limiting process problems. In this case, the initial and final state of a system are specified, and the maximum or minimum amount of heat transfer or work required is obtained. Predicting the maximum output of an ideal gas turbine is an example.
4. Cycle analysis. The analysis of a cyclical sequence of processes, such as the Rankine vapor power cycle.

In fact, each of the above problems is a subset of the first. Consider the adiabatic cylinder shown in Fig. 1.1. A piston separates the cylinder into two subsystems. Several possibilities can occur:

1. The piston is adiabatic, fixed, and impermeable.
2. In the language of thermodynamics, the piston becomes diathermal. This means that heat transfer can occur through the piston.
3. The piston is now allowed to move. Thus, work can take place.
4. The piston becomes porous. Mass transfer is allowed.

The first case means that the system is completely closed to any kind of interaction, and if the subsystems are individually in equilibrium they remain so. Each additional change removes a constraint and results in a possible spontaneous process leading to a new

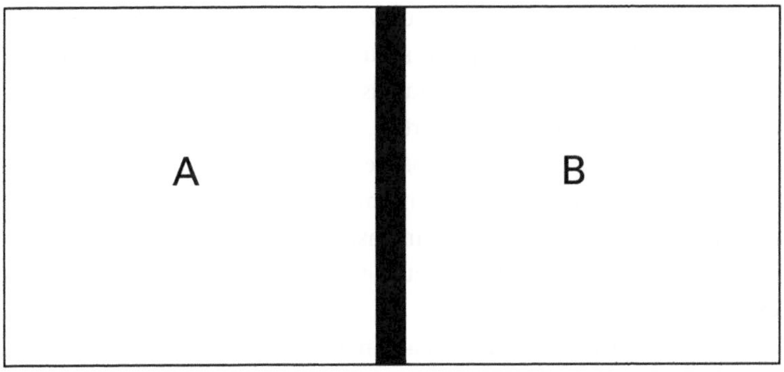

Figure 1.1 The fundamental problem.

equilibrium state. The fundamental problem of thermodynamics is to find the final state once a given constraint is removed. If the fundamental problem can be solved, then type 1–4 problems can also be solved.

Once it is possible to solve the fundamental problem, then all other types of thermodynamic system problems can be solved by breaking the system down into its component processes and analyzing each process individually. This will usually result in a simultaneous set of nonlinear algebraic equations.

1.7 Analysis of Non-equilibrium Behavior

As we have seen, thermodynamics is the study of equilibrium states. That is, given a change in system constraints, what new equilibrium state arises? Thermodynamics does not address the nature of the process or processes that result in a change in equilibrium state. The dynamics part of the word thermodynamics is thus something of a misnomer. Many workers have suggested using the word thermostatics. However, two centuries of usage are not easily put aside, and our use of the name thermodynamics is unlikely to change. More common is to use the term equilibrium thermodynamics.

The question naturally arises, how do we deal with processes and non-equilibrium states? This is the subject of kinetic theory, which leads to the disciplines of fluid mechanics, heat transfer, and mass transfer. As we shall see from our study of microscopic thermodynamics, the equilibrium state arises from the statistical averaging of the possible microscopic states allowed by the macroscopic constraints. In complete forms of kinetic theory equations are derived that describe the departure of these state relations from equilibrium. For example, in a gas there is a unique distribution of atomic velocities that occurs at equilibrium. For momentum transport or heat transfer to occur, this distribution must depart from its equilibrium form. We will explore this subject in a later chapter.

1.8 Summary

In this chapter we explored some of fundamental concepts upon which the field of thermodynamics rests. We started with a short discussion of how matter is composed of atoms and molecules, and that the average motions of these particles at the microscopic level has important implications at the macroscopic level.

Essential to understanding thermodynamics are the concepts of energy, work, heat transfer, and the 1st Law. Work, of course, is a force acting through a distance:

$$W_{12} = \int_{\vec{x}_1}^{\vec{x}_2} \vec{F} \cdot d\vec{x} \tag{1.14}$$

Using Newton's Law $\vec{F} = m\vec{a}$ we derived the concept of kinetic energy as

$$\int_{\vec{x}_1}^{\vec{x}_2} \vec{F} d\vec{x} = \frac{1}{2} m (V_2^2 - V_1^2) \tag{1.15}$$

or

$$KE = \frac{1}{2} m V^2 \tag{1.16}$$

After some manipulation, we also derived the potential energy

$$PE = mgz \tag{1.17}$$

Putting these concepts together results in the 1st Law of Thermodynamics

$$\Delta E = \Delta KE + \Delta PE = W_{12} \tag{1.18}$$

Adding in the possibility of kinetic energy and work taking place at the microscopic level, the 1st Law becomes

$$\Delta E = U + KE + PE = W_{12} + Q_{12} \tag{1.19}$$

Work and heat transfer are processes that depend on the details, or path, by which they take place. Internal, kinetic, and potential energies, on the other hand, are properties related to the amount of work or heat transfer that takes place. In addition, the composition of any working substance is important as well.

Finally, one can cast all thermodynamic problems in terms of the fundamental problem described in Section 1.6. We explore this in detail in Chapter 2.

1.9 Problems

1.1 Compare the kinetic energy of a 0.5-km diameter asteroid approaching the Earth at 30,000 m/sec with the 210,000 TJ released by a 50-megaton hydrogen bomb.

1.2 Calculate the potential energy required for a 160-lbm person to hike to the top of Long's Peak (14,259 ft) from downtown Boulder, CO (5,430 ft). Compare that to the energy in a McDonald's large Big Mac Meal with chocolate shake.

1.3 Estimate the number of molecules in the Earth's atmosphere.

1.4 Calculate the number of molecules in one cubic meter of air at STP. Then estimate the total kinetic energy of this mass assuming that the average speed of molecules is the speed of sound in air at STP. Compare this to the potential energy associated with lifting this mass of air 20 m.

1.5 Consider two equal 1000-cm^3 cubes of copper. Initially separated, one has a temperature of 20 °C and the other is at 100 °C. They are then brought into contact along one wall, but otherwise isolated from their surroundings. Estimate how long it will take for the two cubes to come into equilibrium.

2 Fundamentals of Macroscopic Thermodynamics

In this chapter we explore the basis of macroscopic thermodynamics, introducing the concept of entropy and providing a deeper understanding of properties like temperature, pressure, and chemical potential. We end with a number of property relationships that can be used in practical ways. However, to proceed at the macroscopic level it is necessary to introduce some postulatory basis to our considerations. Clearly, without starting at the microscopic level there is little we can say about the macroscopic averages. Even in statistical mechanics, however, we must make statistical postulates before the mathematical model can be constructed. (Note that we closely follow Callen [3].)

2.1 The Postulates of Macroscopic (Classical) Thermodynamics

The postulates are:

I. **There exist certain states (called equilibrium states) of simple systems that, macroscopically, are characterized completely by the internal energy U, the volume V, and the mole numbers $N_1, N_2, \ldots, N_r$ of the chemical components, where r is the number of chemical components.**

This postulate has the important consequence that previous history plays no role in determining the final state. This implies that all atomic states must be allowed. This is equivalent to making an assumption called a priori equal probability. This assumption is not always met. (Two examples of non-equilibrium states that are quite stable are heat-treated steel and non-Newtonian fluid with hysteresis. We can often live with systems in metastable equilibrium if it does not otherwise interfere with our thermodynamic analysis.)

II. **There exists a function called the entropy, S, of the extensive properties of any composite system, defined for all equilibrium states and having the following property: The values assumed by the extensive properties are those that maximize the entropy over the manifold of constrained equilibrium states.**

This postulate is based on the following argument. Suppose we repeatedly observe a composite system evolve from a given initial state to its final equilibrium state. This is the fundamental problem. If the final state is the same time after time, we would naturally conclude that some systematic functional dependency exists between the initial state of the system and the final state. Mathematically, we would say that the final state is a stable point of this function. That being the case, there must be some function that displays the property of stability at the final state, perhaps a maximum or a minimum. We make that assumption, calling the function entropy, and define it to have a maximum at the final state.

We can solve the basic problem if the relationship between S and the other extensive properties is known for the matter in each subsystem of a thermodynamic composite system. This relationship, called the fundamental relation, is of the form

$$S = S(U, V, N_i) \tag{2.1}$$

and plays a central role in thermodynamics. To evaluate the fundamental relation from first principles requires the application of statistical mechanics, the subject we will study during the second part of this course. As we shall see then, even today it is not always possible to theoretically determine the fundamental relation for complex materials.

It turns out that one could define the entropy in many different ways. The second postulate only asserts that the function be a maximum at the final equilibrium state, and says nothing about other mathematical properties such as linearity, additivity, and so on. The third postulate addresses these issues.

III. The entropy of a composite system is additive over the constituent subsystems. The entropy is continuous and differentiable and is a monotonically increasing function of energy.

This postulate insures that entropy is a well-behaved function, and simplifies the task of finding the final equilibrium state. Additivity assures that the entropy for a composite system can be written as

$$S = \sum_j S_j \tag{2.2}$$

where j is an index identifying subsystems of the composite system. S_j is a function of the properties of the subsystem

$$S_j = S_j(U_j, V_j, N_{ij}) \tag{2.3}$$

A consequence of the additivity property is that the entropy must be a homogeneous, first-order function of the extensive properties

$$S(\lambda U, \lambda V, \lambda N_i) = \lambda S(U, V, N_i) \tag{2.4}$$

where λ is a constant multiplier. This is consistent with the idea that U, V, and N_i are extensive properties, that is, they are a measure of the extent of the system. Thus, entropy must also be an extensive property.

The monotonic property implies that

$$\left.\frac{\partial S}{\partial U}\right)_{V,N_i} > 0 \tag{2.5}$$

The continuity, differentiability, and monotonic properties insure that S can be inverted with respect to U and energy always remains single valued. This allows us to write

$$U = U(S, V, N_i) \tag{2.6}$$

as being equivalent to the entropy form of the fundamental relation. Each form contains all thermodynamic information.

Note that because entropy is extensive, we can write

$$S(U, V, N_i) = NS\left(\frac{U}{N}, \frac{V}{N}, \frac{N_i}{N}\right) \tag{2.7}$$

where N is the total number of moles in the system. For a single-component system, this becomes

$$S(U, V, N_i) = NS\left(\frac{U}{N}, \frac{V}{N}, 1\right) \tag{2.8}$$

If we define as intensified, or normalized properties

$$u \equiv \frac{U}{N}, v \equiv \frac{V}{N}, s \equiv \frac{S}{N} \tag{2.9}$$

then the fundamental relation becomes

$$s = s(u, v) \tag{2.10}$$

This form is most familiar to undergraduate engineering students.

As will be seen later, the final postulate is necessary to determine the absolute value of the entropy.

IV. The entropy of any system vanishes in the state for which

$$\left.\frac{\partial U}{\partial S}\right)_{V,N_j\text{'s}} = 0 \tag{2.11}$$

The fourth postulate is also known as the 3rd Law of Thermodynamics or the Nernst postulate. As we shall see, this postulate states that

$$S \to 0 \text{ as } T \to 0 \tag{2.12}$$

2.2 Simple Forms of the Fundamental Relation

As stated above, the fundamental relation plays a central role in thermodynamic theory. It contains all the necessary information about the equilibrium behavior of specific substances, and is necessary to solve the fundamental problem. For a limited number of cases, analytic forms of the fundamental relations are available. We present two here without derivation to use in subsequent examples.

2.2.1 Van der Waals Substance

In 1873 van der Waals [16] proposed a form for the fundamental relation of a simple compressible substance based on heuristic arguments regarding the nature of intermolecular interactions. His arguments have subsequently been shown to be qualitatively correct, and the resulting fundamental relation is very useful for illustrating typical behaviors.

Van der Waals discovered that when two atoms first approach they experience an attractive force due to the net effect of electrostatic forces between the electrons and the nuclei present. However, if the nuclei are brought into close proximity, the repulsive force of like charges dominates the interaction. The potential energy associated with this force is defined as

$$V(r) = -\frac{dF}{dr} \tag{2.13}$$

and is illustrated in Fig. 2.1. The point of zero force, where the attractive and repulsive forces just balance, corresponds to the minimum in the potential energy. It is this property that allows molecules to form, and subsequently allows atoms and molecules to coalesce into liquids or solids.

Based on these arguments, and utilizing a specific form of the potential energy function, van der Waals showed that one form of the fundamental relation that satisfies the postulates can be written as

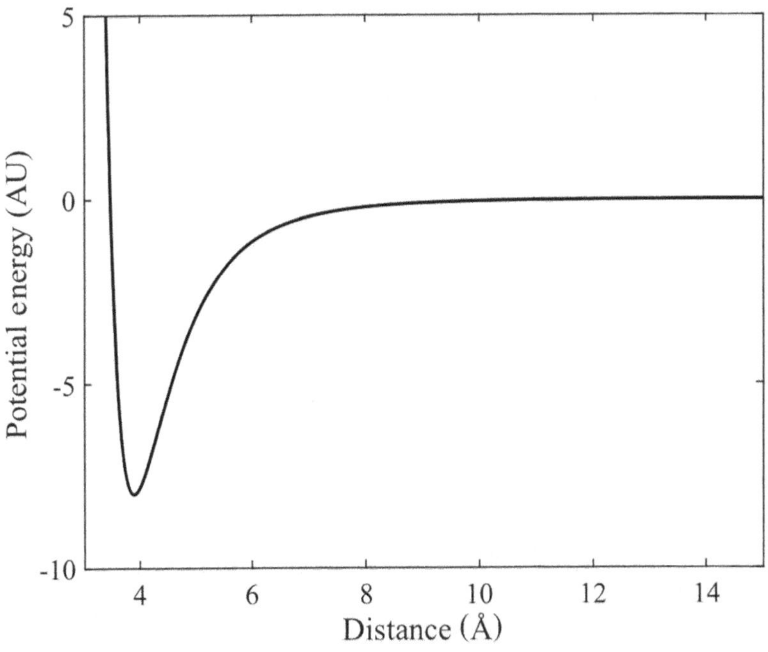

Figure 2.1 Interatomic potential energy for the diatomic molecule.

$$S = NR\left\{\left(\frac{V}{V_0} - b\right)\left(\frac{U}{U_0} + a\frac{V_0}{V}\right)^c\right\} \tag{2.14}$$

where a, b, and c are parameters of the relationship and R is the universal gas constant.

2.2.2 Ideal Gas

In the limit where intermolecular forces play little role in the macroscopic behavior of a substance, V and U become large compared to b and a/v. In this case, called the ideal or dilute gas limit, the fundamental relation becomes

$$S = S_0 + NR\left\{\ln\left(\frac{U}{U_0}\right)^c + \ln\left(\frac{V}{V_0}\right) - (c+1)\ln\left(\frac{N}{N_0}\right)\right\} \tag{2.15}$$

where cR is the specific heat at constant volume and S_0, U_0, and V_0 are reference values of U and V, respectively (i.e. for a monatomic gas $c = 3/2$).

The ideal gas fundamental relation does not satisfy the fourth postulate. This is because it is a limiting expression not valid over the entire range of possible values of the independent parameters.

2.3 Equilibrium and the Intensive Properties

To solve the fundamental problem we follow the dictate of the second postulate. That is, for the composite system we are analyzing we calculate the entropy of each subsystem as a function of the extensive properties. The final values of the unconstrained properties are those that maximize the entropy subject to any system-wide constraints. The maximum of a function with respect to its independent variables is found by setting the total derivative of the function to zero. That is, if

$$S = S(U, V, N_j) \tag{2.16}$$

then

$$dS = \left.\frac{\partial S}{\partial U}\right)_{V,N_i} dU + \left.\frac{\partial S}{\partial V}\right)_{U,N_i} dV + \sum_j \left.\frac{\partial S}{\partial N_j}\right)_{V,N_{i\neq j}} dN_j = 0 \tag{2.17}$$

The partial derivatives, which we have not yet interpreted, are functions of the extensive properties. However, they are in a normalized form, being essentially ratios of extensive properties. As we shall see, they relate directly to the intensive properties we are familiar with.

Of course, setting the derivative to zero does not guarantee that a maximum in entropy has been found. To be certain, the sign of the second derivative must be less than zero, otherwise one may have found a saddle point or a minimum. Furthermore, the overall stability of an equilibrium point will be determined by the character of the entropy function near the point in question. This is the subject of thermodynamic stability and plays an important role in studying phase equilibrium.

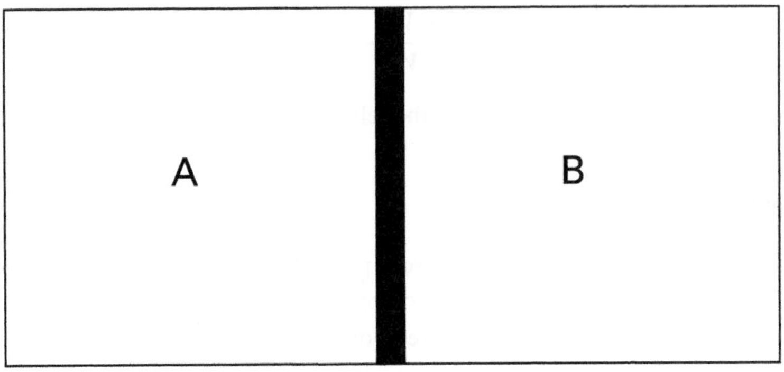

Figure 2.2 Thermal equilibrium.

2.3.1 Thermal Equilibrium: The Meaning of Temperature

Consider the problem of thermal equilibrium, as illustrated in Fig. 2.2. The composite system consists of two adjacent subsystems, thermally isolated from the surroundings and separated by a barrier that is initially an insulator. The problem is to find the final state of the system after the barrier is allowed to conduct heat, or become diathermal. In this problem, only the internal energy can change; both V and N_i are constrained because the barrier is fixed and impervious to mass transfer.

The second postulate would thus have us write

$$S = S_A + S_B \tag{2.18}$$

so that

$$dS = dS_A + dS_B \tag{2.19}$$

or

$$dS = \left.\frac{dS}{dU}\right)_A dU_A + \left.\frac{dS}{dU}\right)_B dU_B \tag{2.20}$$

Now, although the internal energy within each subsystem can change, the overall energy is constrained because the composite system is isolated from its surroundings. Thus,

$$U = U_A + U_B = \text{ const.} \tag{2.21}$$

This is a constraint, or conservation relation. Indeed, this is the statement of the 1st Law of Thermodynamics for this problem, and determines the allowed relationship between U_A and U_B. Thus, we can write

$$dU = 0 = dU_A + dU_B \tag{2.22}$$

and use the result to eliminate dU_B in Eq. (2.20) so that

$$dS = \left\{ \left.\frac{dS}{dU}\right)_A - \left.\frac{dS}{dU}\right)_B \right\} dU_A \tag{2.23}$$

At equilibrium we must have $dS \rightarrow 0$. Since U_A is the independent variable, at equilibrium

$$\left.\frac{dS}{dU}\right)_A = \left.\frac{dS}{dU}\right)_B \tag{2.24}$$

The thermodynamic property we associate with heat transfer and thermal equilibrium is temperature. Heat transfer occurs from a higher-temperature to a lower-temperature body and two systems in thermal equilibrium have equal temperature. Thus, one might be tempted to associate these partial derivatives with temperature. A short thought experiment would prove that were we to do so, however, the heat transfer between two bodies out of equilibrium would be from cold to hot! Therefore, it is conventional to define

$$T \equiv \frac{1}{\left.\frac{\partial S}{\partial U}\right)_{V,N_i}} = \left.\frac{\partial U}{\partial S}\right)_{V,N_i} \tag{2.25}$$

and the equilibrium requirement becomes

$$T_A = T_B \tag{2.26}$$

Temperature and entropy are closely related quantities. Their product must have the dimension of energy, but their individual units are arbitrary. Since the units for temperature were defined empirically long before the concept of entropy was developed, it is conventional to use the historic unit of temperature to define those for entropy. In the SI system the units are Kelvin. Thus the units of entropy must be

$$\left[\frac{J}{K}\right] \tag{2.27}$$

2.3.2 Mechanical Equilibrium: The Meaning of Pressure

Suppose we now replace the barrier separating subsystems A and B in Fig. 2.2 with a diathermal piston. Now, in addition to heat transfer, the volumes can change and work will be done as the piston moves. Unlike the previous problem, both internal energy and volume can change. Therefore, the total derivative of the system entropy must be written

$$dS = \left.\frac{\partial S}{\partial U}\right)_A dU_A + \left.\frac{\partial S}{\partial V}\right)_A dV_A + \left.\frac{\partial S}{\partial U}\right)_B dU_B + \left.\frac{\partial S}{\partial V}\right)_B dV_B \tag{2.28}$$

where it is implicitly assumed that in evaluating the partial derivatives, the other independent property is being held constant.

As in the previous problem, the total energy is constrained, and Eqs (2.21) and (2.22) still hold. In addition, we must consider the constraint on volume

$$V = V_A + V_B = \text{const.} \tag{2.29}$$

that leads to

$$dV = 0 = dV_A + dV_B \tag{2.30}$$

Using the conservation relations of Eqs (2.22) and (2.30) to replace dU_B and dV_B, we get

$$dS = \left\{ \left.\frac{\partial S}{\partial U}\right)_A - \left.\frac{\partial S}{\partial U}\right)_B \right\} dU_A + \left\{ \left.\frac{\partial S}{\partial V}\right)_A - \left.\frac{\partial S}{\partial V}\right)_B \right\} dV_A \tag{2.31}$$

As in the thermal case, dS must go to zero, resulting in the requirement that

$$\left.\frac{\partial S}{\partial U}\right)_A = \left.\frac{\partial S}{\partial U}\right)_B \text{ and } \left.\frac{\partial S}{\partial V}\right)_A = \left.\frac{\partial S}{\partial V}\right)_B \tag{2.32}$$

We have already interpreted the meaning of the first relation, that the temperature of the two subsystems must be equal when there is thermal equilibrium. Likewise, mechanical equilibrium clearly requires that the force on each side of the piston be equal, which in the case of a piston requires that the pressures on each side be equal. Thus, one would not be too surprised to find the partial derivative of entropy with volume to be related to pressure. The dimensions of the derivative must be entropy over volume. However, we set the dimensions of entropy in our discussion on thermal equilibrium, or

$$\begin{aligned}\left[\frac{\text{Entropy}}{\text{Volume}}\right] &= \left[\frac{\text{Energy}}{\text{Volume} * \text{Temperature}}\right] \\ &= \left[\frac{\text{Force} * \text{Length}}{\text{Length}^3 * \text{Temperature}}\right] = \left[\frac{\text{Force}}{\text{Length}^2 * \text{Temperature}}\right]\end{aligned} \tag{2.33}$$

Therefore, the derivative has dimensions of pressure over temperature and we define the thermodynamic pressure with the relationship

$$\frac{p}{T} \equiv \left.\frac{\partial S}{\partial V}\right)_{U,N_i} \tag{2.34}$$

and the full equilibrium requirement becomes

$$T_A = T_B \text{ and } p_A = p_B \tag{2.35}$$

The observant student will wonder why we included heat transfer in the problem of mechanical equilibrium. Interestingly, without heat transfer the problem has no stable solution. Indeed, were the piston released from an initial condition of non-equilibrium without heat transfer (or equivalently friction), it would oscillate forever.

2.3.3 Matter Flow and Chemical Equilibrium: The Meaning of Chemical Potential

Now consider the case where the fixed, diathermal barrier of the first problem becomes permeable to one or more chemical components while remaining impermeable to all others. For this problem internal energy and the mole number (say N_1) of the specie under consideration become the independent variables, and

$$dS = \left.\frac{\partial S}{\partial U}\right)_A dU_A + \left.\frac{\partial S}{\partial N_1}\right)_A dN_{1A} + \left.\frac{\partial S}{\partial U}\right)_B dU_B + \left.\frac{\partial S}{\partial N_1}\right)_B dN_{1B} \tag{2.36}$$

As above we set the derivative to zero, leading to the restriction that

$$\left.\frac{\partial S}{\partial U}\right)_A = \left.\frac{\partial S}{\partial U}\right)_B \text{ and } \left.\frac{\partial S}{\partial N_1}\right)_A = \left.\frac{\partial S}{\partial N_1}\right)_B \tag{2.37}$$

Again, we have already identified the derivative of entropy with internal energy as the inverse of temperature, so that as before equilibrium requires that the subsystem temperatures be equal.

The dimensions of $\partial S/\partial N$ are entropy over mole number, or

$$\left[\frac{\text{Entropy}}{\text{Moles}}\right] = \left[\frac{\text{Energy}}{\text{Mole} * \text{Temperature}}\right] \tag{2.38}$$

It is conventional to identify the derivative as

$$\frac{\mu_1}{T} \equiv -\left.\frac{\partial S}{\partial N_1}\right)_{U,V,N_{i\neq 1}} \tag{2.39}$$

where μ is called the chemical potential and plays the same role as pressure in determining a state of equilibrium with respect to matter flow. That is, the subsystem chemical potentials must be equal at equilibrium for each specie involved in mass transfer, and we have

$$T_A = T_B \text{ and } \mu_A = \mu_B \tag{2.40}$$

This reasoning can be applied to chemically reacting systems as follows. Consider the case of a closed system, whose overall energy and volume are fixed, but within which an initial batch of r chemical compounds is allowed to chemically react. In this case, the entropy is a function only of the mole numbers of the reacting species. Therefore

$$dS = \sum_{i=1}^{r} \frac{\mu_i}{T} dN_i \tag{2.41}$$

Setting this expression to zero results in one equation with r unknown mole numbers. For a chemically reacting system, atoms are conserved, not moles. The mole numbers are only constrained in a way that insures conservation of atoms. If n_{ki} is the number of k-type atoms in species i, and p_k is the number of moles of k-type atoms initially present in the system, then

$$p_k = \sum_{i=1}^{r} n_{ki} N_i \tag{2.42}$$

This is a equations, where a is the number of different types of atoms within the system. If the fundamental relation is known, then these equations may be solved for the unknown equilibrium mole numbers using the method of undetermined multipliers as described in Chapter 5 for ideal gas mixtures.

2.4 Representation and the Equations of State

We have been writing the functional form of the fundamental relation as

$$S = S(U, V, N_i) \tag{2.43}$$

This form is called the entropy representation. Based on the considerations of the previous section, we can now write the differential form of this relation as

Table 2.1 Equations of state

Van der Waals fluid	Ideal gas
$\frac{1}{T} = \frac{cR}{u+a/v}$	$\frac{1}{T} = \frac{cR}{u}$
$\frac{p}{T} = \frac{acR}{uv^2+av}$	$\frac{p}{T} = \frac{R}{v}$

$$dS = \frac{1}{T}dU + \frac{p}{T}dV - \sum_{i=1}^{r} \frac{\mu_i}{T}dN_i \tag{2.44}$$

We could also write

$$U = U(S, V, N_i) \tag{2.45}$$

Because of the statement regarding the monotonic relationship between entropy and energy in the third postulate, this relation, called the energy representation, contains precisely the same information as its expression in the entropy representation. It is, therefore, also a form of the fundamental relation. The differential form can be written as

$$dU = TdS - pdV + \sum_{i=1}^{r} \mu_i dN_i \tag{2.46}$$

In both representations the coefficients in the differential form are intensive parameters that can be calculated as a function of the independent properties from the fundamental relation. Such relationships are generally called equations of state.

In the energy representation, for example, the equations of state take on the functional forms

$$\begin{aligned} T &= T(S, V, N_i) \\ p &= p(S, V, N_i) \\ \mu_i &= \mu_i(S, V, N_i) \end{aligned} \tag{2.47}$$

Note that if the equations of state are known, then the differential form of the fundamental relation can be recovered by integration. The equations of state in the energy representation for a single-component van der Waals fluid and ideal gas are given in Table 2.1.

2.5 The Euler Equation and the Gibbs–Duhem Relation

There are two very useful general relationships that can be derived directly using the properties of the fundamental relation. Called the Euler equation and the Gibbs–Duhem relation, they relate the extensive and intensive properties in particularly direct ways. Starting from the homogeneous, first-order property, we can write, for any λ

$$U(\lambda S, \lambda V, \lambda N_i) = \lambda U(S, V, N_i) \tag{2.48}$$

Differentiating with respect to λ,

$$\frac{\partial U}{\partial(\lambda S)}\frac{\partial(\lambda S)}{\partial\lambda}+\frac{\partial U}{\partial(\lambda V)}\frac{\partial(\lambda V)}{\partial\lambda}+\sum_{i=1}^{r}\frac{\partial U}{\partial(\lambda N_i)}\frac{\partial(\lambda N_i)}{\partial\lambda}=U \tag{2.49}$$

Noting that $\frac{\partial(\lambda S)}{\partial\lambda}=S$, and so on, and taking $\lambda=1$,

$$\frac{\partial U}{\partial S}S+\frac{\partial U}{\partial V}V+\sum_{i=1}^{r}\frac{\partial U}{\partial N_i}N_i=U \tag{2.50}$$

or

$$U=TS-pV+\sum_{i=1}^{r}\mu_i N_i \tag{2.51}$$

This is called the Euler equation. In the entropy representation it can be written

$$S=\frac{1}{T}U+\frac{p}{T}V-\sum_{i=1}^{r}\frac{\mu_i}{T}N_i \tag{2.52}$$

The Gibbs–Duhem relation can be derived as follows. Start with the energy representation Euler equation in differential form:

$$dU=TdS+SdT-pdV-Vdp+\sum_{i=1}^{r}\mu_i dN_i+\sum_{i=1}^{r}N_i d\mu_i \tag{2.53}$$

However, the differential form of the fundamental relation (energy representation) is

$$dU=TdS-pdV+\sum_{i=1}^{r}\mu_i dN_i \tag{2.54}$$

Subtracting the two, we obtain the Gibbs–Duhem relation

$$SdT-Vdp+\sum_{i=0}^{r}N_i d\mu_i=0 \tag{2.55}$$

The Gibbs–Duhem relation shows that not all the intensive parameters are independent of each other. The actual number of independent intensive parameters is called the thermodynamic degrees of freedom. A simple system of r components thus has $r+1$ degrees of freedom. As we shall see, this corresponds directly to Gibbs' phase rule.

2.6 Quasi-static Processes and Thermal and Mechanical Energy Reservoirs

Many thermodynamic systems of interest interact with surroundings that are much larger than the system under study and whose properties therefore change minimally upon interaction with the system. Examples include heat exchange with a large body of water or the atmosphere, and expansion against the atmosphere. Idealizing this interaction simplifies system analysis. Part of our idealization will be that during interaction with a reservoir, the reservoir evolves in a quasi-static manner. By quasi-static we mean that the processes occur slowly enough that the reservoir can be treated as though it were in equilibrium at all times, and in the case of the mechanical energy reservoir described below, the process is frictionless. Some texts use the term quasi-equilibrium to describe quasi-static processes.

The quasi-static idealization is very important in thermodynamic theory as it forms the basis of reversibility, a concept we will use to analyze the limiting behavior of systems.

First consider the mechanical energy reservoir (MER) represented in Fig. 2.3 as a very large piston and cylinder without friction. We use this reservoir to model work done on (or by) the surroundings. If a system does work on the MER, the 1st Law can be written

$$dU = \delta W \tag{2.56}$$

If the work is carried out in a quasi-static manner, then

$$dS = \frac{1}{T}dU + \frac{p}{T}dV \tag{2.57}$$

Since $\delta W = -pdV$, this leads to the MER being isentropic.

The thermal energy reservoir (TER) is illustrated in Fig. 2.4. We use this reservoir to model heat transfer with the surroundings. If a system transfers heat to the TER, the 1st Law for the TER becomes

$$dU = \delta Q \tag{2.58}$$

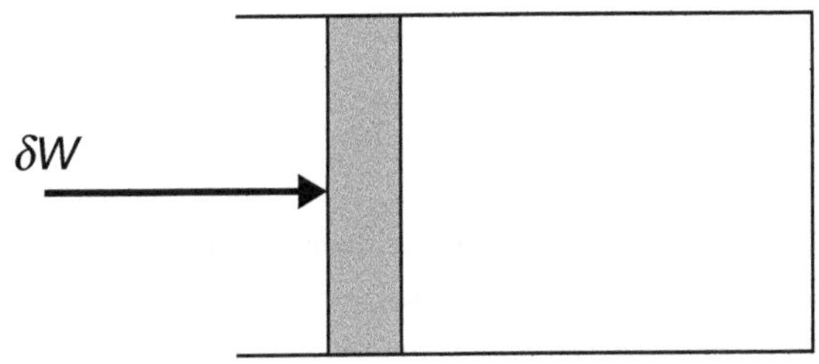

Figure 2.3 Mechanical energy reservoir.

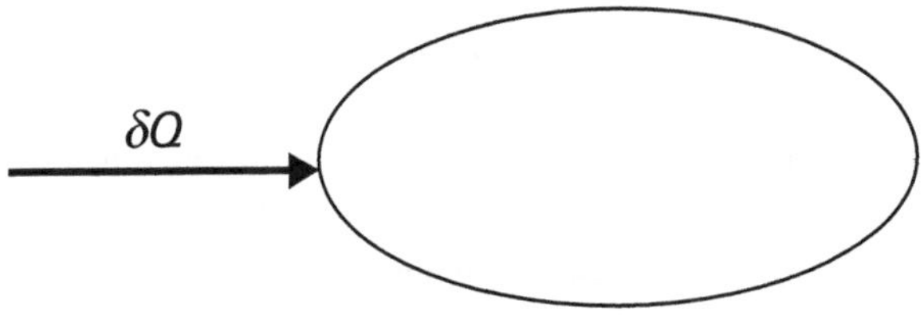

Figure 2.4 Thermal energy reservoir.

For this constant-volume system, the differential form of the fundamental relation becomes

$$dS = \frac{1}{T} dU \tag{2.59}$$

Thus, the entropy of a TER is not constant. In fact, by the 1st Law, $dU = \delta Q$, so that

$$dS = \frac{\delta Q}{T} \tag{2.60}$$

This is Clausius' famous relationship with which he defined entropy.

2.7 Equilibrium in the Energy Representation

In thermodynamic systems analysis we are very much interested in understanding the relationship between independent and dependent variables. In the entropy representation, U, V, and N_i are the independent variables while S, T, p, and μ_i are the dependent variables. The fundamental problem is to determine the values of the dependent properties in each subsystem given that certain processes are allowed. In the entropy representation the properties are determined by maximizing the entropy of the overall system.

The energy representation offers a different set of independent variables, namely S, V, and N_i. Since S is not a conserved quantity, thermal and mechanical energy reservoirs are required to control the independent properties in the energy representation. It is logical to ask how the final values of the dependent properties in the subsystems are determined. In fact, the equilibrium state is determined by minimizing the energy of the system (not including the reservoirs), a result that will be important in considering other combinations of dependent and independent properties.

To prove the energy minimization principle, consider a system that is allowed to interact with reservoirs. The second postulate (2nd Law) holds for the system and the reservoirs. Now, assume that the energy of the system (not including the reservoirs) does not have the smallest possible value consistent with a given total entropy. Then withdraw energy from the system as work into a MER, maintaining the entropy of the system constant. From the fundamental relation,

$$dS = \frac{\delta Q}{T} \tag{2.61}$$

Thus, since S is constant,

$$dU = -pdV \tag{2.62}$$

which is the work done on the MER. Now, return the system to its original energy by transferring heat from a TER. However, from the fundamental relation,

$$dU = TdS \tag{2.63}$$

Since the temperature is positive, S must have increased. Thus, the original state could not have been in equilibrium.

2.8 Alternative Representations – Legendre Transformations

In practice, there are several possible combinations of dependent and independent properties. We have focused so far on using the fundamental relation in the entropy representation and touched briefly on the energy representation. However, there are a variety of thermodynamic problems for which these representations are not appropriate. In many problems we may wish to treat one of the intensive properties as an independent variable. Because the intensive properties are derivatives of the extensive properties, however, rearranging the fundamental relation must be carried out with care using the method of Legendre transformations (see [17]).

Consider the case of a fundamental relation with a single independent variable,

$$Y = Y(X) \tag{2.64}$$

The intensive property is

$$z = \frac{\partial Y}{\partial X} \tag{2.65}$$

We wish to obtain a suitable function with z as the independent variable that contains all the information of the original function. We could replace X with z between these two relationships, so that

$$Y = Y(z) \tag{2.66}$$

However, the replacement would only be accurate to within an integration constant. In other words, to invert the relationship we must integrate $z(Y)$ to recover X, or

$$X = \int \frac{dY}{z} + C \tag{2.67}$$

where C is unknown.

A method which takes this problem into account is the Legendre transformation. As illustrated in Fig. 2.5, it involves defining the transformed relationship in terms of both the local slope, z, and the Y intercept, Ψ. For a given value of X,

$$z = \frac{Y - \Psi}{X - 0} \tag{2.68}$$

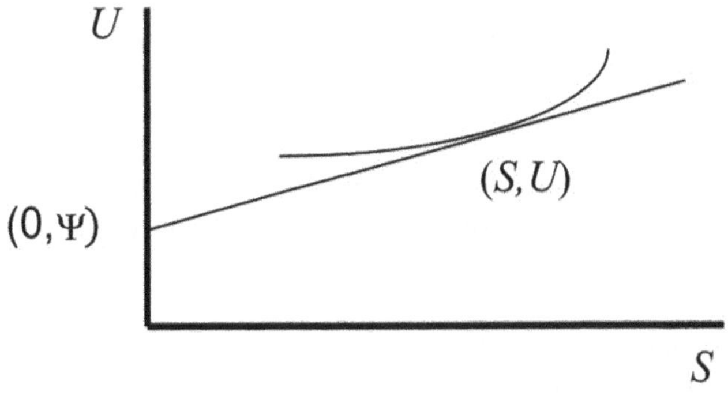

Figure 2.5 Graphical illustration of the transformation process.

thus

$$\Psi = Y - zX \tag{2.69}$$

The procedure is to use Eqs (2.68) and (2.69) to eliminate X and Y in Eq. (2.66).

2.8.1 Example 2.1

Perform the forward and reverse Legendre transformation for

$$Y = X^2 \qquad [2.8.1]$$

Then

$$z = 2X$$

Thus

$$Y = \left(\frac{z}{2}\right)^2 = \frac{z^2}{4}$$

which with

$$X = \frac{z}{2}$$

leads to

$$\Psi = \frac{z^2}{4} - \frac{z^2}{2} = -\frac{z^2}{4}$$

For the reverse problem we are given

$$\Psi = \Psi(z)$$

We wish to recover

$$Y = Y(X)$$

Since

$$\Psi = Y - zX$$

we can write

$$d\Psi = dY - zdX - Xdz$$

but

$$dY = zdX$$

so that

$$d\Psi = -Xdz \quad \text{or} \quad -X = \frac{d\Psi}{dz}$$

Use this relation and Eq. [2.8.1] to eliminate z and Ψ. Using our example,

$$-X = -\frac{z}{2}$$

thus

$$Y = X^2$$

We can generalize by writing the form of the fundamental relation as

$$Y = Y(X_0, X_1, \ldots, X_t) \tag{2.70}$$

where Y is the entropy or energy and X_i are the appropriate extensive properties. The intensive properties are then

$$z_k \equiv \left.\frac{\partial Y}{\partial X_k}\right)_{X_{i \neq k}} \tag{2.71}$$

The transformation procedure then involves fitting planes, instead of straight lines, with

$$\Psi = Y - \sum_k z_k X_k \tag{2.72}$$

where the summation is over the independent properties we wish to replace. To invert Ψ, take the total derivative

$$d\Psi = dY - \sum_k z_k dX_k - \sum_k X_k d_k = -\sum_k X_k dz_k \tag{2.73}$$

Thus

$$-X_k = \left.\frac{\partial \Psi}{\partial z_k}\right)_{P_{i \neq k}} \tag{2.74}$$

Use this and Eq. (2.70) to replace Ψ and z_R.

Note that we don't need to replace every independent variable, only those that are convenient.

2.9 Transformations of the Energy

We now consider transformations of the energy representation. These functions are called thermodynamic potential functions because they can be used to predict the ability of given systems to accomplish work, heat transfer, or mass transfer.

In the energy representation the z_k of the previous section correspond to T, $-p$, and μ_i. For a simple compressible substance, in addition to energy there are seven possible transformations as listed in Table 2.2.

In the table, the notation $U[\;]$ is used for those functions which do not have an established name. (The term "canonical" was introduced by Gibbs to describe different system/reservoir combinations that he analyzed using ensemble statistical mechanics. We will explore these combinations in Chapter 5.)

2.10 Transformations of the Entropy

The transformations of entropy are of more practical use in using statistical mechanics to determine the fundamental relation. In general, these transforms are called Massieu functions, and the bracket notation $S[\;]$ is used to identify them. They are listed in Table 2.3.

Table 2.2 Transformations of the energy representation

Function name	Form	Derivative
Energy	$U = U(S, V, N)$	$dU = TdS - pdV + \mu dN$
Helmholtz	$F = F(T, V, N) = U - TS$	$dF = -SdT - pdV + \mu dN$
Enthalpy	$H = H(S, p, N) = U + pV$	$dH = TdS + Vdp + \mu dN$
	$U[S, V, \mu] = U - \mu N$	$dU[S, V, \mu] = TdS - pdV - Nd\mu$
Gibbs	$G = G(T, p, N) = U - TS + pV$	$dG = -SdT + Vdp + \mu dN$
Grand canonical	$U[T, V, \mu] = U - TS - \mu N$	$dU[T, V, \mu] = -SdT - pdV - Nd\mu$
	$U[S, p, \mu] = U + pV - \mu N$	$dU[S, p, \mu] = -SdT - pdV + \mu dN$
Microcanonical	$U[T, p, \mu] =$ $U - TS + pV - \mu N = 0$	$dU[T, p, \mu] = -SdT + Vdp - Nd\mu = 0$

Table 2.3 Transformations of the entropy representation

Function name	Form	Derivative
Entropy	$S = S(U, V, N)$	$dS = \frac{1}{T}dU + \frac{p}{T}dV - \frac{\mu}{T}dN$
Canonical	$S[1/T, V, N] = S - \frac{1}{T}U$	$dS[1/T, V, N] = -Ud1/T + \frac{p}{T}dV - \frac{\mu}{T}dN$
	$S[U, p/T, N) = S - \frac{p}{T}V$	$dS[U, p/T, N] = \frac{1}{T}dU - Vd\frac{p}{T} - \frac{\mu}{T}dN$
	$S[U, V, \mu/T] = S + \frac{\mu}{T}N$	$dS[U, V, \mu/T] = \frac{1}{T}dU + \frac{p}{T}dV + Nd\frac{\mu}{T}$
	$S[1/T, p/T, N] =$ $S - \frac{1}{T}U - \frac{p}{T}V$	$dS[1/T, p/T, N] = -Ud\frac{1}{T} - Vd\frac{p}{T} - \mu dN$
Grand canonical	$S[1/T, V, \mu/T] =$ $S - \frac{1}{T}U + \frac{\mu}{T}N$	$dS[1/T, V, \mu/T] = -Ud\frac{1}{T} + \frac{p}{T}dV + Nd\frac{\mu}{T}$
	$S[U, p/T, \mu/T] =$ $S - \frac{p}{T}V + \frac{\mu}{T}N$	$dS[U, p/T, \mu/T] = \frac{1}{T}dU - Vd\frac{p}{T} + Nd\frac{\mu}{T}$
Microcanonical	$S[1/T, p/T, \mu/T] =$ $S - \frac{1}{T}U - \frac{p}{T}V + \frac{\mu}{T}N = 0$	$dS[1/T, p/T, \mu/T] = -Ud\frac{1}{T} - Vd\frac{p}{T} + Nd\frac{\mu}{T} = 0$

2.11 Reversible Work

The transformations of energy are particularly useful in the following way. Suppose we ask what is the maximum amount of work that can be done by a system that is held at a constant temperature (Fig. 2.6).

In particular, we assume that all processes occur in a quasi-static fashion. Then, by the 1st Law,

$$U_{SYS} + U_{TER} + U_{MER} = U_{TOT} \tag{2.75}$$

or

$$dU_{MER} = -(dU_{SYS} + dU_{TER}) \tag{2.76}$$

However, from our discussion of reservoirs,

$$dU_{TER} = -TdS \tag{2.77}$$

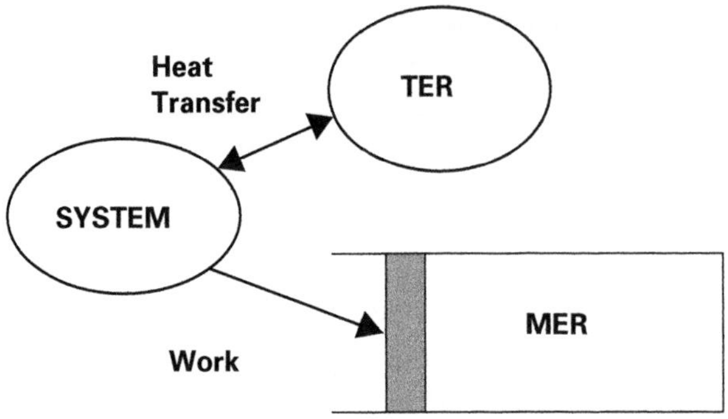

Figure 2.6 Constant-temperature work.

Therefore,

$$dU_{MER} = -dF \tag{2.78}$$

Thus, the Helmholtz function is a measure of the energy available to do useful work by a constant-temperature system. For this reason the Helmholtz function is also called the Helmholtz free energy.

Likewise, enthalpy is the measure of energy available to do reversible work by a constant-pressure system, or

$$dU_{MER} = -dH \tag{2.79}$$

The Gibbs function is the measure of energy available to do reversible work by a constant-temperature and constant-pressure system, or

$$dU_{MER} = -dG \tag{2.80}$$

One can show that each of the thermodynamic potentials is a measure of a system to perform reversible work when the indicated intensive parameters are held constant.

2.12 Maxwell's Relations

Maxwell [18] derived a set of relationships based on the properties of differentiation that relate the various intensive and extensive properties. The relations are particularly helpful in utilizing experimental tabular data.

Consider the general function of two variables

$$f = f(x, y) \tag{2.81}$$

By the chain rule of differentiation

$$df = \left.\frac{\partial f}{\partial x}\right)_y dx + \left.\frac{\partial f}{\partial y}\right)_x dy = Mdx + Ndy \tag{2.82}$$

Now

$$\left.\frac{\partial M}{\partial y}\right)_x = \frac{\partial f}{\partial x \partial y} \text{ and } \left.\frac{\partial N}{\partial x}\right)_y = \frac{\partial f}{\partial y \partial x} \tag{2.83}$$

but

$$\frac{\partial f}{\partial x \partial y} = \frac{\partial f}{\partial y \partial x} \tag{2.84}$$

so that

$$\left.\frac{\partial M}{\partial y}\right)_x = \left.\frac{\partial N}{\partial x}\right)_y \tag{2.85}$$

Applying this result to the energy of a simple compressible substance, we note that M and N correspond to T and $-p$ in the energy representation. In other words,

$$U = U(S, V) \tag{2.86}$$

$$dU = TdS - pdV \tag{2.87}$$

Therefore,

$$\left.\frac{\partial T}{\partial V}\right)_S = \left.\frac{\partial(-p)}{\partial S}\right)_V \tag{2.88}$$

Likewise, from $dF = -SdT - pdV$,

$$\left.\frac{\partial(-S)}{\partial V}\right)_T = \left.\frac{\partial(-p)}{\partial T}\right)_V \tag{2.89}$$

from $dH = TdS + VdP$,

$$\left.\frac{\partial T}{\partial p}\right)_S = \left.\frac{\partial V}{\partial S}\right)_p \tag{2.90}$$

and from $dG = -SdT + Vdp$,

$$\left.\frac{\partial T}{\partial p}\right)_S = \left.\frac{\partial V}{\partial S}\right)_p \tag{2.91}$$

2.13 Building Property Relations

In practice, we seek properties as a function of the independent variables, not just the fundamental relation. We can use the various transforms discussed above and Maxwell's relations to assist us in doing so. Let us start by considering the practical case of using laboratory data to construct relations. In the laboratory only certain of the extensive and intensive properties are amenable to direct measurement. These include temperature, pressure, volume, mole number, and force. For a simple compressible substance, the relationship between the first four provides the pvT relationship. By measuring the force necessary to change the volume, relative energy changes can be measured, which in turn allow determination of the specific heats. As we shall see, given the pvT relation and the specific heats, the fundamental relation can be determined. Once the fundamental relation is known in one representation, it can be transformed to any other representation.

Consider the differential form of the Massieu function for a simple compressible substance:

$$dS[1/T, V, N] = -Ud(1/T) + \frac{p}{T}dV - \frac{\mu}{T}dN \tag{2.92}$$

If we know the $pVTN$ relationship, then the second term is determined, and it remains to determine the energy and chemical potential. The normalized form of the energy can be written as

$$u = u(T, v) \tag{2.93}$$

The total derivative for this expression is

$$du = \left.\frac{\partial u}{\partial T}\right)_v dT + \left.\frac{\partial u}{\partial v}\right)_T dv \tag{2.94}$$

The first term is the specific heat at constant volume,

$$c_v = \left.\frac{\partial u}{\partial T}\right)_v \tag{2.95}$$

which can be measured. It remains to interpret the second term. Consider the normalized form of the fundamental relations in the energy representation

$$du = Tds - pdv \tag{2.96}$$

Express entropy as a function of temperature and specific volume, $s = s(T, v)$, then

$$ds = \left.\frac{\partial s}{\partial T}\right)_v dT + \left.\frac{\partial s}{\partial v}\right)_T dv \tag{2.97}$$

Inserting this into Eq. (2.96),

$$du = T\left.\frac{\partial s}{\partial T}\right)_v dT + \left[T\left.\frac{\partial s}{\partial v}\right)_T - p\right]dv \tag{2.98}$$

The coefficient of the first term must be the specific heat at constant volume, so that

$$c_v = T\left.\frac{\partial s}{\partial T}\right)_v \tag{2.99}$$

From Maxwell's relations,

$$\left.\frac{\partial s}{\partial v}\right)_T = \left.\frac{\partial p}{\partial T}\right)_v \tag{2.100}$$

so that

$$du = c_v dT + \left[T\left.\frac{\partial p}{\partial T}\right)_v - p\right]dv \tag{2.101}$$

This is the first term in the Massieu function.

The chemical potential can be obtained from the Gibbs–Duhem relation, which in the entropy representation is

$$d\mu = T\left(ud\frac{1}{T} + vd\frac{p}{T}\right) \tag{2.102}$$

Hence, by supplying c_v and the pvT relation we have recovered the fundamental relation in a Massieu transformation form.

By invoking Maxwell's relations a number of other useful relations can be obtained, including

$$c_p - c_v = T\left(\frac{\partial v}{\partial T}\right)_p \left(\frac{\partial p}{\partial T}\right)_v \tag{2.103}$$

$$c_p - c_v = -T\left(\frac{\partial v}{\partial T}\right)_p^2 \left(\frac{\partial p}{\partial v}\right)_T \tag{2.104}$$

It is common to define the coefficients of thermal expansion and isothermal compressibility as

$$\alpha \equiv \frac{1}{v}\left(\frac{\partial v}{\partial T}\right)_p \tag{2.105}$$

and

$$\kappa_T \equiv -\frac{1}{v}\left(\frac{\partial v}{\partial p}\right)_T \tag{2.106}$$

Then

$$c_p - c_v = \frac{vT\beta^2}{\alpha} \tag{2.107}$$

The Joule–Thompson coefficient [19] is defined as

$$\mu \equiv \left(\frac{\partial T}{\partial p}\right)_h \tag{2.108}$$

and describes how the temperature changes in adiabatic throttling:

$$\mu \begin{cases} < 0 \text{ temperature increases} \\ \phantom{<}\ 0 \text{ temperature constant} \\ > 0 \text{ temperature decreases} \end{cases}$$

This is illustrated in Fig. 2.7 for several substances.

To determine the fundamental relation from first principles will require statistical methods as discussed in the Introduction. That is the subject of the following chapters.

2.14 Sources for Thermodynamic Properties

There are numerous sources for thermodynamic properties. Data is contained in textbooks, and in various handbooks. One can also find numerous open source and commercial property programs. Developing and compiling thermodynamic data is part of

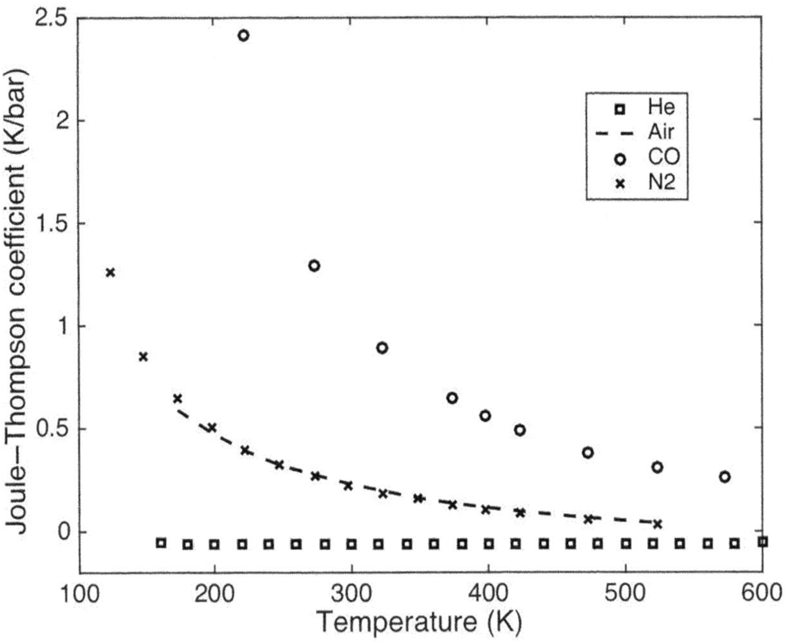

Figure 2.7 Joule–Thompson coefficient for several substances (data from *Perry's Chemical Engineers' Handbook* [8]).

NIST's mission. Their program REFPROP [20] (https://www.nist.gov/srd/refprop) is a good compilation of data for a variety of substances, particularly refrigerants.

2.15 Summary

2.15.1 Postulates and the Fundamental Relation

We started by outlining the postulates of classical thermodynamics:

I. There exist certain states (called equilibrium states) of a simple system that, macroscopically, are characterized completely by the internal energy U, the volume V, and the mole numbers $N_1, N_2, \ldots, N_r$ of the chemical components, where r is the number of chemical components.

II. There exists a function called the entropy, S, of the extensive parameters of any composite system, defined for all equilibrium states and having the following property: The values assumed by the extensive parameters are those that maximize the entropy over the manifold of constrained equilibrium states.

III. The entropy of a composite system is additive over the constituent subsystems. The entropy is continuous and differentiable and is a monotonically increasing function of energy.

IV. The entropy of any system vanishes in the state for which

$$\left.\frac{\partial U}{\partial S}\right)_{V,N_j\text{'s}} = 0 \tag{2.109}$$

From these postulates arises the concept of the fundamental relation

$$S = S(U, V, N_i) \tag{2.110}$$

Examples of fundamental relations are those for a van der Waals substance and an ideal gas:

$$s = s_0 + R\ln\left\{(v-b)\,(u + a/v)^c\right\} \tag{2.111}$$

and

$$s = s_0 + cR\ln\frac{u}{u_0} + R\ln\frac{v}{v_0} \tag{2.112}$$

2.15.2 Equilibrium and Intensive Parameters

We then derived the thermodynamic definitions of temperature

$$T \equiv \frac{1}{\left.\frac{\partial S}{\partial U}\right)_{V,N_i}} = \left.\frac{\partial U}{\partial S}\right)_{V,N_i} \tag{2.113}$$

pressure

$$\frac{p}{T} \equiv \left.\frac{\partial S}{\partial V}\right)_{U,N_i} \tag{2.114}$$

and chemical potential

$$\frac{\mu_1}{T} \equiv -\left.\frac{\partial S}{\partial N_1}\right)_{U,V,N_{i\neq 1}} \tag{2.115}$$

2.15.3 Representation and Equations of State

Representation refers to the idea that once the fundamental relation is known, then by suitable transformation dependent and independent variables can be exchanged for convenience. For example, the third postulate allows us to invert energy and entropy in Eq. (2.110) resulting in

$$U = U(S, V, N_i) \tag{2.116}$$

We say that this is the energy representation and Eq. (2.110) is in the entropy representation.

Both equations can be differentiated and take on the form

$$dS = \frac{1}{T}dU + \frac{p}{T}dV - \sum_{i=1}^{r}\frac{\mu_i}{T}dN_i \tag{2.117}$$

and

$$dU = TdS - pdV + \sum_{i=1}^{r} \mu_i dN_i \tag{2.118}$$

In both representations the coefficients in the differential form are intensive parameters that can be calculated as a function of the independent properties from the fundamental relation. Such relationships are generally called equations of state. In the energy representation, for example, the equations of state take on the functional forms

$$\begin{aligned} T &= T(S, V, N_i) \\ p &= p(S, V, N_i) \\ \mu_i &= \mu_i(S, V, N_i) \end{aligned} \tag{2.119}$$

2.15.4 The Euler Equation and the Gibbs–Duhem Relation

Two useful relations we derived are the Euler relation

$$U = TS - pV + \sum_{i=1}^{r} \mu_i N_i \tag{2.120}$$

and the Gibbs–Duhem relation

$$SdT - Vdp + \sum_{i=0}^{r} N_i d\mu_i = 0 \tag{2.121}$$

2.15.5 Alternative Representations

By performing Legendre transformations, we showed that we could exchange dependent and independent variables in either of the entropy of energy representations. This led to the Massieu functions and thermodynamic potentials, respectively. The transforms are given in Tables 2.2 and 2.3.

2.15.6 Maxwell's Relations

A useful set of relations can be derived by noting that the cross derivatives of the coefficients of a derivative relation are equal. This leads to the following four relationships, known as Maxwell's relations:

$$\left.\frac{\partial T}{\partial V}\right)_S = \left.\frac{\partial(-p)}{\partial S}\right)_V \tag{2.122}$$

$$\left.\frac{\partial(-S)}{\partial V}\right)_T = \left.\frac{\partial(-p)}{\partial T}\right)_V \tag{2.123}$$

$$\left.\frac{\partial T}{\partial p}\right)_S = \left.\frac{\partial V}{\partial S}\right)_p \tag{2.124}$$

$$\left.\frac{\partial T}{\partial p}\right)_S = \left.\frac{\partial V}{\partial S}\right)_p \tag{2.125}$$

2.15.7 Property Relations

By utilizing the various relationships we have derived, we can also derive various practical property relations. Typically we require two types of relations for practical problem solving; a pvT relation and a caloric equation of state. The pvT relation is obtained either from experiment, or by calculation from first principles using the methods of statistical thermodynamics. The caloric equation of state is most easily obtained from a relation like Eq. (2.126):

$$du = c_v dT + \left[T \left. \frac{\partial p}{\partial T} \right)_v - p \right] dv \tag{2.126}$$

where the specific heat is either measured or calculated. The chemical potential can likewise be found using

$$d\mu = T \left(u d\frac{1}{T} + v d\frac{p}{T} \right) \tag{2.127}$$

2.16 Problems

2.1 Fundamental equations can come in various functional forms. Consider the following equations and check to see which ones satisfy Postulates II, III, and IV. For each, functionally sketch the relation between S and U. (In cases where there are fractional exponents, take the positive root.)

(a)
$$S \sim (NVU)^{2/3}$$

(b)
$$S \sim \frac{NU}{V}^{1/3}$$

(c)
$$U \sim \frac{S^3}{V^2} \exp\left(S/NR\right)$$

(d)
$$U \sim (UV)^2 \left(1 + \frac{S}{NR}\right) \exp\left(-S/NR\right)$$

2.2 Two systems of monatomic ideal gases are separated by a diathermal wall. In system A there are 2 moles initially at 175 K, in system B there are 3 moles initially at 400 K. Find U_A, U_B as a function of R and the common temperature after equilibrium is reached.

2.3 Two systems have these equations of state:

$$\frac{1}{T_1} = \frac{3}{2} R \frac{N_1}{U_1}$$

and

$$\frac{1}{T_2} = \frac{5}{2} R \frac{N_2}{U_2}$$

where R is the gas constant per mole. The mole numbers are $N_1 = 3$ and $N_2 = 5$. The initial temperatures are $T_1 = 175$ K and $T_2 = 400$ K. What are the values of U_1 and U_2 after equilibrium is established?

2.4 Two systems have the following equations of state:

$$\frac{1}{T_1} = \frac{3}{2}R\frac{N_1}{U_1}, \qquad \frac{P_1}{T_1} = R\frac{N_1}{V_1}$$

and

$$\frac{1}{T_2} = \frac{5}{2}R\frac{N_2}{U_2}, \qquad \frac{P_2}{T_2} = R\frac{N_2}{V_2}$$

The two systems are contained in a closed cylinder, separated by a fixed, adiabatic and impermeable piston. $N_1 = 2.0$ and $N_2 = 1.5$ moles. The initial temperatures are $T_1 = 175$ K and $T_2 = 400$ K. The total volume is 0.025 m^3. The piston is allowed to move and heat transfer is allowed across the piston. What are the final energies, volumes, temperatures, and pressures once equilibrium is reached?

2.5 Find the three equations of state for a system with the fundamental relation

$$U = C\frac{S^4}{NV^2}$$

Show that the equations of state are homogeneous zero order, that is T, P, and μ are intensive parameters. (C is a positive constant.)

2.6 A system obeys the following two equations of state:

$$T = \frac{2As}{v^2}$$

and

$$P = \frac{2As^2}{v^3}$$

where A is a constant.
Find μ as a function of s and v, and then find the fundamental equation.

2.7 Find the three equations of state for the simple ideal gas (see Table 2.1) and show that these equations of state satisfy the Euler relation.

2.8 Find the relation between the volume and the temperature of an ideal van der Waals fluid in an isentropic expansion.

2.9 Find the fundamental relation for a monatomic gas in the Helmholtz, enthalpy, and Gibbs representations. Start with Eq. (2.15) and assume a monatomic gas with $c_v = 3/2$.

2.10 A system obeys the fundamental relation

$$(s - s_0)^4 = Avu^2$$

Find the Gibbs potential $G(T, P, N)$.

2.11 Prove that c_v must vanish for any substance as T goes to zero.

2.12 Find the Joule–Thompson coefficient for a gas that obeys

$$V = RT/P + aT^2 \text{ and } \qquad c_p = A + BT + CP$$

where a, A, B, C, and R are constants.

2.13 What is temperature?

2.14 What is pressure?

2.15 What is chemical potential?

2.16 What is entropy?

3 Microscopic Thermodynamics

Our next great task is determining the form of the fundamental relation based on the microscopic nature of matter. As we shall see, that task reduces to relating the macroscopic independent properties to a distribution of microscopic quantum states that characterizes the equilibrium macroscopic state. In the next section, we provide background and explore the plausibility of the postulates. We then present two postulates that prescribe how we are to proceed. From then on we explore the consequences of the postulates and provide a relationship between the microscopic and macroscopic scales.

3.1 The Role of Statistics in Thermodynamics

In the Introduction, we discussed some aspects of the problem. We know that matter is composed of atoms and molecules. The size of an atom is determined by the range of electrostatic forces due to the positively charged nucleus and negatively charged electrons, and is approximately 5 Å or 5×10^{-10} m. There are about 2×10^{9} atoms per lineal meter in a solid, resulting in about 8×10^{27} atoms/m^3. In a gas at standard conditions, there are about 10^{24} atoms/m^3. Thus, macroscopic systems are composed of a very, very large number of atoms or molecules. Even if we ignore the fact that the behavior of individual atoms and molecules is governed by quantum mechanics, it would be impossible to simultaneously solve the enormous number of equations of motion involved to describe the evolution of a macroscopic system to equilibrium. Clearly, an alternative approach is required and fortunately, nature has been kind in devising the laws of averaging in ways that allow for great simplification.

Consider the following thought experiment. Suppose N atoms are placed in a box of fixed volume V, which is then closed and adiabatic. N is chosen small enough so that the system is always in the gas phase. The kinetic energy ϵ of each particle is the same and nonzero, and when the particles are put in the box they are all placed in one corner. Only kinetic energy is considered and for simplicity, we will assume that the trajectory of each particle is described by Newton's Law. How will this system evolve?

We know by conservation of energy that the average kinetic energy of the particles must remain the same no matter how the overall dynamic state of the system evolves. However, because the particles have kinetic energy they will move within the volume, beginning to distribute themselves around the available space. They must remain within the volume, so they might collide with the wall. They might also collide with each other.

Because the number of particles is very large, it would not take very much time before it would be practically impossible to reconstruct the initial placement of the particles. (For mathematicians, this leads to a chaotic state.) Finally, after some period of time that is long compared to diffusional and convective times, we might expect that the particles would be roughly evenly distributed throughout the volume.

In fact, the above description is exactly what would happen. We call the final state the equilibrium state. Were we to measure the local number and energy densities (looking at a volume that is small compared to the dimensions of the box but large compared to the average molecular spacing), we would find them to be uniform throughout the volume and with time, even though the particles continue to move, colliding with each other and the walls. We would also find the positions and velocities of individual particles to be highly randomized. (We can actually make such a measurement using laser scattering to confirm this behavior.)

We know from quantum mechanics that, in the absence of applied forces, the dynamic state of an atom or molecule will relax to a fixed quantum state. The atom or molecule arrives at that state as a result of interactions with other particles. The state is constantly changing, because interactions occur at a very high frequency, 10^9 or 10^{10}/sec at room temperature and pressure. Indeed, it is through intermolecular interactions that equilibrium is reached. At any instant in time, the complete specification of the quantum state of every atom or molecule describes the quantum state of the entire macroscopic system. Thus, the system quantum state is continually changing as well. However, the system is subject to the macroscopic constraints, fixed numbers, and energy for a closed system, for example. Therefore, the system quantum state is constrained.

Suppose we know the system quantum state, meaning that we know exactly what quantum state each atom or molecule is in. In that case we could calculate the system value of any mechanical property by summing the value of that property for each particle. However, we have painted the following picture. At equilibrium, it appears at the macroscopic level that the local mechanical properties of the system are constant. When a macroscopic constraint is changed, the system evolves to a new equilibrium state with new mechanical properties. However, even when a system is in equilibrium, the system quantum state is constantly changing. Therefore, what system quantum state could we use to characterize the equilibrium system?

The number of system quantum states that can satisfy a given set of macroscopic constraints is enormous, approximately e^N where N is the number of particles (see Appendix E in Knuth [5]). Some of these physically allowed states cannot possibly be equilibrium states. For example, in our thought experiment above, the initial state with all the particles in one corner of the box satisfies the macroscopic constraints, yet we clearly will not actually observe that state for more than a few microseconds. Equilibrium always occurs when the constraints remain fixed for a sufficient period of time. Therefore, most system quantum states must look very similar (i.e. states that result in equilibrium behavior must be overwhelmingly more probable than other states).

We might be tempted to make the following argument. The system quantum state is constantly changing. No individual allowed system quantum state is physically preferred over any other state. (This is called equal a priori probability.) If we were to wait long enough, we would see every allowed state. Therefore, equilibrium is really the average over all the allowed states. In fact, this argument is incorrect. While every state that satisfies the macroscopic constraints is physically allowable, dynamics precludes that a system will be able to evolve through all its allowed states in a reasonable time. For example, the collision frequency in a gas at standard conditions is approximately 10^{10}/sec. For a 1-m^3 system, it would take about $e^{10^{34}}$ sec to access every allowed state once. This time is longer than the age of the universe! This argument reaffirms the conclusion from above that allowed quantum states which look like equilibrium must be overwhelmingly more probable than states that do not.

If equilibrium quantum states are so much more probable than non-equilibrium ones, then one way to proceed is to look at the relative probability of each allowed system quantum state, and characterize equilibrium with the most probable one. Indeed, that is the approach we will follow. The only remaining question is how to find that state. There are two historical approaches. The first makes the equal a priori assumption discussed above. It involves watching a single system and finding the most probable of all the allowed states. The difficulty with this approach is that it has no physical analog, it being physically impossible for all allowed states to appear in a single system.

The second approach we will take is that followed by Gibbs [4]. It involves the following argument. Imagine a very large number of identical macroscopic systems. Gibbs called this collection an ensemble, and each individual system a member of the ensemble (Fig. 3.1). If the number of ensemble members is large enough (which it can be, since we are imagining the ensemble), then at any instant in time all allowed quantum states will appear.

There are two major advantages of this concept. The first, and most obvious, is that we need not make the equal a priori approximation. Rather than observe over impossibly large times, we take a snapshot at a fixed time. The second advantage is that we can construct the ensemble to mimic real systems. That is, we can include a reservoir. The only requirement is that the reservoir be very large, so that all appropriate intensive properties are held constant for every ensemble member. Of course, we will have to construct the ensemble so that all members experience the same conditions. All constraints and allowed interactions with the reservoir are the same. We will also treat the entire ensemble as closed.

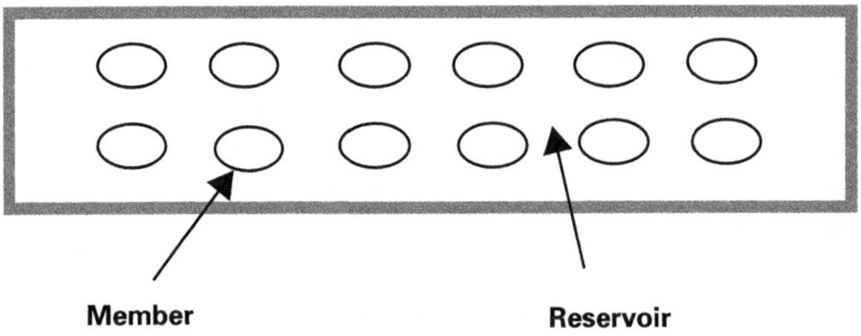

Figure 3.1 An ensemble of ensemble members.

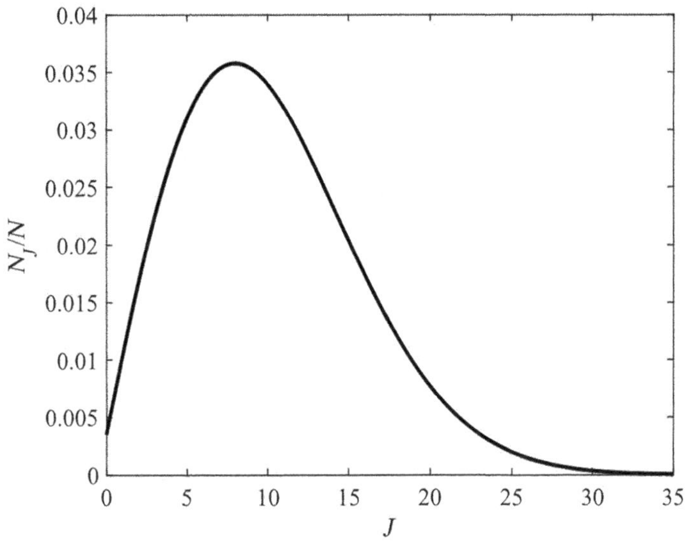

Figure 3.2 Equilibrium rotational population distribution for CO.

Suppose we are able to find the most probable system quantum state for a given set of independent properties. How would we characterize the state and what could we do with it? Recall that the quantum state of a macroscopic system is specified by the distribution of microscopic quantum states. By distribution, we mean a normalized count of how many particles are presently existing in each possible quantum state. For example, Fig. 3.2 shows a distribution of rotational energy for a diatomic gas at standard conditions plotted as a function of a rotational energy index, J. From the plot we can find the fraction of molecules that have a given rotational energy. There will be similar distributions for translational, vibrational, and electronic energy.

Suppose we index all possible system quantum states with the index j. Then, if n_j is the number of ensemble members that exist in system quantum state j, and n is the number of ensemble members, we can calculate the mechanical parameters U, V, and N_i:

$$\langle A \rangle = \frac{1}{n} \sum_j n_j A_j \tag{3.1}$$

The bracket indicates average or expectation value, and

j = ensemble member quantum state index
$A_j = U_j$, V_j, or N_{ij}
n_j = number of ensemble members in quantum state j
n = total number of ensemble members

Note that the set of n_j's is the quantum state distribution for the ensemble. Many different sets are possible; our task is to find the most probable.

3.2 The Postulates of Microscopic Thermodynamics

The postulates we will invoke were first stated by Gibbs in 1902. In the foreword of his book he stated, “The only error into which one can fall, is the want of agreement between the premises and the conclusions, and this, with care, one may hope, in the main, to avoid.” It turns out that he was right, and his formulation forms the backbone of modern statistical thermodynamics.

The postulates are:

I. **The macroscopic value of a mechanical–thermodynamic property of a system in equilibrium is characterized by the expected value computed from the most probable distribution of ensemble members among the allowed ensemble member quantum states.**

II. **The most probable distribution of ensemble members among ensemble member quantum states is that distribution for which the number of possible quantum states of the ensemble and reservoir is a maximum.**

Postulate I states that we can calculate $\langle A \rangle$ using

$$\langle A \rangle = \frac{1}{n}\sum_j n_j^* A_j = \sum_j p_j A_j \tag{3.2}$$

where $p_j = n_j^*/n$ is the normalized probability distribution and * indicates that we are using the most probable distribution of n_j's or p_j's.

Postulate II describes how to determine the most probable distribution. To apply the postulate we must obtain an expression for the number of allowed quantum states as a function of the n_j. Then, find the n_j that maximizes this function.

For an ensemble involving a reservoir, the members are in some quantum state, the reservoir in another. If Ω is the number of states available to the members, and Ω_R the number available to the reservoir, then the total number available to the entire ensemble including the reservoir is

$$\Omega_{TOT} = \Omega_R \Omega \tag{3.3}$$

This is the joint probability of two independent random processes.

It will turn out that we don't need Ω_R. We will need, however, an expression for Ω in terms of n_j. The question is: How many ways (permutations) can n members in a given distribution (combinations) of quantum states be arranged? From basic probability mathematics,

$$\Omega = \frac{n!}{\prod_j n_j!} \tag{3.4}$$

Thus,

$$\Omega_{TOT} = \Omega_R \frac{n!}{\prod_j n_j!} \tag{3.5}$$

We wish to maximize this function, subject to any appropriate constraints. It will turn out to be easier if we maximize the natural log of the function, applying Sterling's approximation [21] to simplify the algebra. Stirling's approximation is (you can test this yourselves)

$$\ln(x!) = x\ln x - x, \text{ if } x >> 1 \tag{3.6}$$

Taking the logarithm of Ω_{TOT} and applying the approximation, we obtain

$$\ln \Omega_{TOT} = \ln \Omega_R + n\ln n - \sum_j n_j \ln n_j \tag{3.7}$$

3.3 The Partition Function and its Alternative Formulations

To proceed further, we need to make a decision about what type of ensemble to use. The possible choices are shown in Table 3.1. Note the correspondence with the macroscopic Massieu functions. We will start with the grand canonical representation. Thus, only the volume of the ensemble members is fixed, and we will need constraints on energy and number of particles. We also need to recognize that there is a constraint on the sum of all the n_j which must equal n. Thus, the constraints become

$$n = \sum_j n_j \tag{3.8}$$

$$U_{TOT} = U_R + \sum_j n_j U_j \tag{3.9}$$

$$N_{iTOT} = N_{iR} + \sum_j n_j N_{ij} \tag{3.10}$$

where U_j and N_{ij} are, respectively, the ensemble member energies and number of i-type particles associated with the ensemble member quantum state j. (These will eventually be determined using quantum mechanics.)

We will use Lagrange's method of undetermined multipliers (see Courant and Hilbert [22]) to maximize Ω_{TOT} subject to the constraints. We start by taking the total derivative of $\ln \Omega_{TOT}$ and setting it to zero:

Table 3.1 Types of ensembles

Gibbs' name	Independent parameters	Fundamental relation
Microcanonical	U, V, N_i	$S(U, V, N_i)$
Canonical	$1/T, V, N_i$	$S[1/T]$
	$U, V, -\mu_i/T$	$S[-\mu_i/T]$
Grand canonical	$1/T, V, -\mu_i/T$	$S[1/T, V, -\mu_i/T]$
	$U, p/T, N_i$	$S[U, p/T, N_i]$
	$1/T, p/T, N_i$	$S[1/T, p/T, N_i]$
	$U, p/T, -\mu_i/T$	$S[U, p/T, -\mu_i/T]$
	$1/T, p/T, -\mu_i/T$	$S[1/T, p/T, -\mu_i/T]$

$$d \ln \Omega_{TOT} = \frac{\partial \ln \Omega_R}{\partial U_R} dU_R + \sum_i \frac{\partial \ln \Omega_R}{\partial N_{iR}} dN_{iR} - \sum_j (1 + \ln n_j) dn_j = 0 \tag{3.11}$$

Since n, U_{TOT}, and N_{iTOT} are constants, by differentiation we obtain

$$0 = \sum_j dn_j \tag{3.12}$$

$$0 = dU_R + \sum_j U_j dn_j \tag{3.13}$$

$$0 = dN_{iR} + \sum_j N_{ij} dn_j \tag{3.14}$$

Since these derivatives are each equal to zero, there is no reason why each cannot be multiplied by a constant factor, nor why they can't be added to or subtracted from the expression for $d \ln \Omega_{TOT}$. Historically, the multipliers chosen are $\alpha - 1$, β, and γ_i and they are subtracted from Eq. (3.11). Doing so:

$$\left[\beta - \frac{\partial \ln \Omega_R}{\partial U_R}\right] dU_R + \sum_i \left[\gamma_i - \frac{\partial \ln \Omega_R}{\partial N_{iR}}\right] dN_{iR}$$
$$+ \sum_j \left(\ln n_j + \alpha + \beta U_j + \sum_i \gamma_i N_{ij}\right) dn_j = 0 \tag{3.15}$$

Since U_R, n_{iR}, and n_j are independent variables, the coefficients of their derivatives must each be zero. Thus,

$$\beta = \left(\frac{\partial \ln \Omega_R}{\partial U_R}\right)_{N_{iR}, V} \tag{3.16}$$

$$\gamma_i = \left(\frac{\partial \ln \Omega_R}{\partial N_{iR}}\right)_{U_R, N_{kR, k \neq i}, V} \tag{3.17}$$

and

$$\alpha = -\ln n_j - \beta U_j - \sum_i \gamma_i N_{ij} \tag{3.18}$$

The most probable distribution, the n_j^*'s, can be found from the equation for α:

$$n_j^* = \exp\left(-\alpha - \beta U_j - \sum_i \gamma_i N_{ij}\right) \tag{3.19}$$

If we sum over n_j^*, then

$$n = e^{-\alpha} \sum_j e^{-\beta U_j - \sum_i \gamma_i N_{ij}} \tag{3.20}$$

or

$$e^{\alpha} = \frac{\sum_j e^{-\beta U_j - \sum_i \gamma_i N_{ij}}}{n} \tag{3.21}$$

and thus

$$p_j = \frac{n_j^*}{n} = \frac{e^{-\beta U_j - \sum_i \gamma_i N_{ij}}}{\sum_j e^{-\beta U_j - \sum_i \gamma_i N_{ij}}} \tag{3.22}$$

The denominator of this expression plays a very important role in thermodynamics. It is called the partition function, in this case the grand partition function:

$$Q_G(\beta, \gamma_i) \equiv \sum_j e^{-\beta U_j - \sum_i \gamma_i N_{ij}} \tag{3.23}$$

The physical significance of the partition function is that it determines how ensemble members are distributed, or partitioned, over the allowed quantum states. Also, note that α has dropped out of the expression for both p_j and Q_G.

The Lagrange multipliers, β and γ_i, are associated with U_j and N_{ij}, respectively. It would not be too surprising if they turned out to be related to $1/T$ and $-\mu_i/T$. However, the product βU_j must be unitless, as must $\gamma_i N_{ij}$. This can be accomplished by dividing by Boltzmann's constant, k, the gas constant per molecule, so that

$$\beta = \frac{1}{kT} \tag{3.24}$$

$$\gamma_i = \frac{-\mu_i}{kT} \tag{3.25}$$

By carrying out an analysis similar to that above for all the possible representations, we would find that in each case, the probability distribution can be represented in the form

$$p_j = \frac{e^{-f_j}}{\sum_j e^{-f_j}} \tag{3.26}$$

where the form of f_j depends on the representation. The denominator is, in each case, identified as the partition function. Furthermore, one can show that the natural logarithm of the partition function times Boltzmann's constant is the fundamental relation in the form of entropy or its associated Massieu function:

$$S[\,] = k \ln Q \tag{3.27}$$

This result is tabulated in Table 3.2.

3.4 Thermodynamic Properties

It remains to interpret the multipliers and relate what we have learned to the fundamental relation. To do so, assume that we are working with the canonical representation, so that

$$Q = \sum_j \exp(-\beta U_j) \tag{3.28}$$

Table 3.2 Types of partition functions

Gibbs' name	Independent parameters	Partition function	Fundamental relation
Microcanonical	U, V, N_i	$Q = \Omega$	$S(U, V, N_i)$
Canonical	$1/T, V, N_i$	$Q = \sum_j \exp\left(-\frac{U_j}{kT}\right)$	$S[1/T]$
	$U, V, -\mu_i/T$	$Q = \sum_j \exp\left(\sum_i \frac{\mu_i}{kT} N_{ij}\right)$	$S[-\mu_i/T]$
Grand canonical	$1/T, V, -\mu_i/T$	$Q = \sum_j \exp\left(-\frac{U_j}{kT} + \sum_i \frac{\mu_i}{kT} N_{ij}\right)$	$S[1/T, V, -\mu_i/T]$
	$U, p/T, N_i$	$Q = \sum_j \exp\left(-\frac{P}{kT} V_j\right)$	$S[U, p/T, N_i]$
	$1/T, p/T, N_i$	$Q = \sum_j \exp\left(-\frac{U_j}{kT} - \frac{P}{kT} V_j\right)$	$S[1/T, p/T, N_i]$
	$U, p/T, -\mu_i/T$	$Q = \sum_j \exp\left(-\frac{P}{kT} V_j + \sum_i \frac{\mu_i}{kT} N_{ij}\right)$	$S[U, p/T, -\mu_i/T]$
	$1/T, p/T, -\mu_i/T$	$Q = \sum_j \exp\left(-\frac{U_j}{kT} - \frac{P}{kT} V_j + \sum_i \frac{\mu_i}{kT} N_{ij}\right)$	$S[1/T, p/T, -\mu_i/T]$

and

$$p_j = \frac{e^{-\beta U_j}}{Q} \tag{3.29}$$

Apply the first postulate to calculate the energy,

$$\langle U \rangle = \sum p_j U_j \tag{3.30}$$

Taking the derivative of U,

$$dU = \sum_j U_j dp_j + \sum p_j dU_j \tag{3.31}$$

The first term can be treated by noting that, by using the definition of p_j, we can express U_j as

$$U_j = \frac{1}{\beta}\left(\ln p_j + \ln Q\right) \tag{3.32}$$

Also,

$$\sum_j \ln p_j dp_j = d\left(\sum_j p_j \ln p_j\right) \tag{3.33}$$

Thus,

$$\sum_j U_j dp_j = -\frac{1}{\beta} d\left(\sum_j p_j \ln p_j\right) \tag{3.34}$$

The second term can be simplified by introducing the concept of ensemble member pressure. Recall that pressure is the derivative of internal energy with respect to volume. For a single j this becomes (here we will temporarily use capital P for pressure to avoid confusion with the probabilities p_j)

$$P_j = -\frac{\partial U_j}{\partial V} \tag{3.35}$$

Therefore,

$$P = \sum_j p_j P_j = -\sum_j p_j \frac{\partial U_j}{\partial V} \tag{3.36}$$

Utilizing these results,

$$dU = -\frac{1}{\beta} d\left(\sum_j p_j \ln p_j\right) - PdV \tag{3.37}$$

Compare this result with the differential relation

$$dU = TdS - PdV \tag{3.38}$$

If we assume that

$$\beta = \frac{1}{kT} \tag{3.39}$$

then

$$S = -k \sum_j p_j \ln p_j \tag{3.40}$$

If we substitute the expression for p_j into this result for S, then

$$S = \frac{1}{T} U + k \ln Q \tag{3.41}$$

Compare this with the definition of $S[1/T]$

$$S\left[\frac{1}{T}\right] = S - \frac{1}{T} U \tag{3.42}$$

thus

$$S\left[\frac{1}{T}\right] = k \ln Q \tag{3.43}$$

One can show that these results apply for any representation.

The function

$$I = -\sum_j p_j \ln p_j \tag{3.44}$$

is known as the "information" in information theory. One can show that I is maximized when all the p_j are equal or all states are equally probable. Thus, if we start in one quantum state, then the natural tendency is to evolve in such a way as to pass through many other states. Equilibrium means that things have evolved long enough to "forget" the initial conditions and access distributions that look like the most probable distribution. Of course, in thermodynamics the distribution is constrained, but the principle remains.

One way to interpret the importance of entropy is to study the effect of quasi-static work or heat transfer on the entropy. From the above, recall that

$$\langle U \rangle = \sum p_j U_j \tag{3.45}$$

and

$$dU = \sum_j U_j dp_j + \sum p_j dU_j \tag{3.46}$$

The second term on the right was shown to be equivalent to $-PdV$. Therefore,

$$dU = \sum_j U_j dp_j - PdV \tag{3.47}$$

Comparing this to the 1st Law,

$$dU = \delta Q + \delta W \tag{3.48}$$

If we assume that the work is reversible, then we can make the associations

$$\delta Q = \sum_j U_j dp_j \tag{3.49}$$

$$\delta W = \sum_j p_j dU_j \tag{3.50}$$

For work to occur, only the U_j's are required to change, not the p_j's. If this occurs, then the entropy is constant and we call the work isentropic. However, for there to be heat transfer, the p_j's must change, and thus the entropy changes.

Note that reversible work is possible only when the allowed quantum states are allowed to change, by changing the volume for example. Otherwise, the only way to change the energy is by heat transfer.

3.5 Fluctuations

Having found the means to obtain the fundamental relation based on the most probable distribution, let us now turn to asking what level of fluctuations we might expect in real systems under equilibrium conditions. A measure of the width of the distribution function is the variance, which is defined as

$$\sigma^2(A) \equiv \left\langle (A - \langle A \rangle)^2 \right\rangle \tag{3.51}$$

Plugging in the expression for the expectation value of a mechanical variable, this can be written as

$$\sigma^2(A) = \sum_j p_j A^2 - \langle A \rangle^2 \tag{3.52}$$

Since we know the p_j, we can calculate the variance. Consider, for example, the variance of energy in the canonical representation:

$$\sigma^2(U) = \sum_j p_j U_j^2 - \langle U \rangle^2 \tag{3.53}$$

This can be written as

$$\sigma^2(U) = \frac{\sum_j U_j^2 e^{-\beta U_j}}{\sum_j e^{-\beta U_j}} - \left\{ \frac{\sum_j U_j e^{-\beta U_j}}{\sum_j e^{-\beta U_j}} \right\}^2 \tag{3.54}$$

or

$$\sigma^2(U) = -\left\{ \frac{\partial}{\partial \beta} \frac{\sum_j U_j e^{-\beta U_j}}{\sum_j e^{-\beta U_j}} \right\} = \left(\frac{\partial^2 \ln Q}{\partial \beta^2} \right) \tag{3.55}$$

However,

$$\langle U \rangle = -\frac{\partial \ln Q}{\partial \beta} \tag{3.56}$$

so that

$$\sigma^2(U) = -\left(\frac{\partial^2 U}{\partial \beta^2} \right)_{V,N} = \frac{1}{kT^2} \left(\frac{\partial^2 U}{\partial T^2} \right)_{V,N} \tag{3.57}$$

For an ideal gas the caloric equation of state is

$$U = c_v NT \tag{3.58}$$

so that

$$\sigma^2(U) = c_v NkT^2 \tag{3.59}$$

and

$$\frac{\sigma(U)}{U} = \sqrt{\frac{k}{c_v N}} \tag{3.60}$$

Since $c_v = \frac{3}{2}k$ for a monatomic gas, k and c_v are of the same order. Clearly, fluctuations only become important when the number of molecules is very small.

This development can be generalized for any mechanical property that is allowed to fluctuate in any representation:

$$\sigma^2(A) = \frac{\partial^2 \ln Q}{\partial a^2} \tag{3.61}$$

where a is the associated multiplier for A, that is β, π, or γ_i. (π is the usual symbol for p/kT.)

3.6 Systems with Negligible Inter-particle Forces

If a system is composed of particles that are indistinguishable from one another, and do not, for the vast majority of time, interact with one another, then for all practical purposes the mechanical properties of the system are those of each particle, summed over all the particles.

We start with some definitions:

U_j = energy of ensemble member in quantum state j
N_j = number of particles in ensemble member in quantum state j
U_{ij} = energy associated with type-i particle of member in quantum state j
N_{ij} = number of i-type particles in member in quantum state j
ϵ_{ik} = energy of i-type particle in particle quantum state k
N_{ikj} = number of i-type particles in particle quantum state k in member in quantum state j
i = particle type index
j = member quantum state number
k = particle quantum state number
(Note: when discussing statistics, N usually means number rather than moles.)

In Chapter 4 we will see that, using quantum mechanics, we can obtain N_{ij} and ϵ_{ij} so that we are now in a position to start calculating the mechanical properties and the partition function. In the grand canonical representation, we are concerned with the ensemble member energy and particle numbers, which become:

$$U_j = \sum_i U_{ij} = \sum_i \sum_k \epsilon_{ik} N_{ikj} \tag{3.62}$$

$$N_j = \sum_i N_{ij} = \sum_i \sum_k N_{ikj} \tag{3.63}$$

The partition function then becomes

$$\begin{aligned} Q &= \sum_j \exp(-\beta U_j - \sum_i \gamma_i N_{ij}) \\ &= \sum_j \exp(-\sum_i \sum_k (\beta \epsilon_{ik} + \gamma_i) N_{ikj}) \\ &= \sum_j \prod_i \prod_k \exp(-(\beta \epsilon_{ik} + \gamma_i) N_{ikj}) \end{aligned} \tag{3.64}$$

There is a simplification of the final form of Eq. (3.64) that is quite useful:

$$Q = \prod_i \prod_k \sum_{\eta=0}^{\text{Max } N_{ikj}} e^{-(\beta \epsilon_{ik} + \gamma_i)\eta} \tag{3.65}$$

where Max N_{ikj} is the maximum number of i-type particles which may simultaneously occupy the kth particle quantum state. The advantage of this form for the partition function is that the sum is for a single quantum state of a single-type particle. We will shortly express this sum in closed algebraic form, even without specifying the exact functional form for ϵ_{ik}. (Derivation of this form is left to a homework problem.)

To proceed further, we must take note of a physical phenomenon that can have a profound impact on the final form of the partition function. For some types of particles, it can be shown that within a given ensemble member, no two particles can occupy the same quantum state. This is called the Pauli exclusion principle (see Pauling and Wilson [23]) and is based on the symmetry properties of the wave functions of the particles being considered. It is found that for some indistinguishable particles, the sign of the wave function changes if the coordinates of the particles are reversed. If so, the function is anti-symmetric, otherwise it is symmetric. For particles with anti-symmetric wave functions, it is also found that only one particle can occupy a given quantum state at one time. This is the exclusion principle.

Particles are labeled according to their symmetry properties:

Symmetry	Max N_{ikj}	Particle name	System name	Examples
Anti-symmetric	1	fermion	Fermi–Dirac	D, He^3, e
Symmetric	infinity	boson	Bose–Einstein	H, He^4, photons

One may show that the partition function for the Fermi–Dirac and Bose–Einstein cases becomes:

$$Q_{FD,BE} = \prod_i \prod_k (1 \pm e^{-\beta\epsilon_{ik}-\gamma_i})^{\pm 1} \tag{3.66}$$

$$\ln Q_{FD,BE} = \sum_i \sum_k \ln (1 \pm e^{-\beta\epsilon_{ik}-\gamma_i})^{\pm 1} \tag{3.67}$$

Note that we can now evaluate N_i, since

$$\langle N_i \rangle = -\left.\frac{\partial \ln Q}{\partial \gamma_i}\right)_{\beta,V,\gamma\neq i} \tag{3.68}$$

so that

$$\langle N_i \rangle_{FD,BE} = \sum_k \frac{1}{e^{+\beta\epsilon_{ik}+\gamma_i} \pm 1} \tag{3.69}$$

Likewise,

$$\langle N_{i,k} \rangle_{FD,BE} = \frac{1}{e^{+\beta\epsilon_{ik}+\gamma_i} \pm 1} \tag{3.70}$$

A special case exists when $e^{+\beta\epsilon_{ik}+\gamma_i} >> 1$. Called the Maxwell–Boltzmann limit, it arises when the density is so low that the issue of symmetry is no longer important. In this case:

$$\ln Q_{MB} = \sum_i \sum_k e^{-\beta\epsilon_{ik}-\gamma_i} \tag{3.71}$$

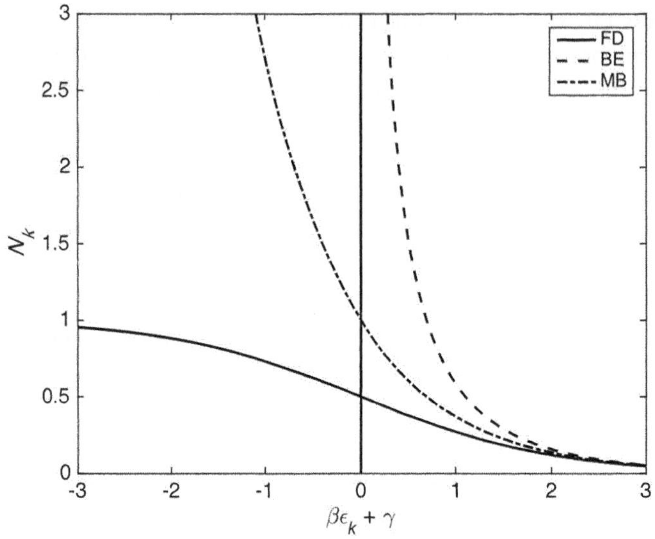

Figure 3.3 Expected value of $< N_k >$.

$$\langle N_i \rangle_{MB} = \sum_k e^{-\beta\epsilon_{ik} - \gamma_i} \tag{3.72}$$

$$\langle N \rangle_{MB} = \sum_i < N_i > = \ln Q_{MB} \tag{3.73}$$

The Maxwell–Boltzmann limit is important because it leads to the ideal gas law. Recall that since

$$k \ln Q = \frac{PV}{T} \tag{3.74}$$

we must have

$$PV = NkT \tag{3.75}$$

This is actually the proof that $\beta = \frac{1}{kT}$.

For those instances where the Maxwell–Boltzmann limit does not apply, the thermodynamic behavior is strongly influenced by whether the particles are bosons or fermions. We can illustrate this point by plotting $\langle N_k \rangle$ as a function of $\beta\epsilon_k + \gamma$ in Fig. 3.3. Note that for a Fermi–Dirac system, $\langle N_k \rangle$ can never exceed unity. For a Bose–Einstein system, $\langle N_k \rangle$ can exceed unity for small values of $\beta\epsilon_k + \gamma$. One may interpret this difference as an effective attraction between bosons, and an effective repulsion between fermions. For large values of $\beta\epsilon_k + \gamma$, the Maxwell–Boltzmann limit holds and inter-particle forces are irrelevant.

The effect of temperature and chemical potential on the distribution of particles among the particle quantum states for Bose–Einstein and Fermi–Dirac systems is illustrated in Figs 3.4 and 3.5, respectively.

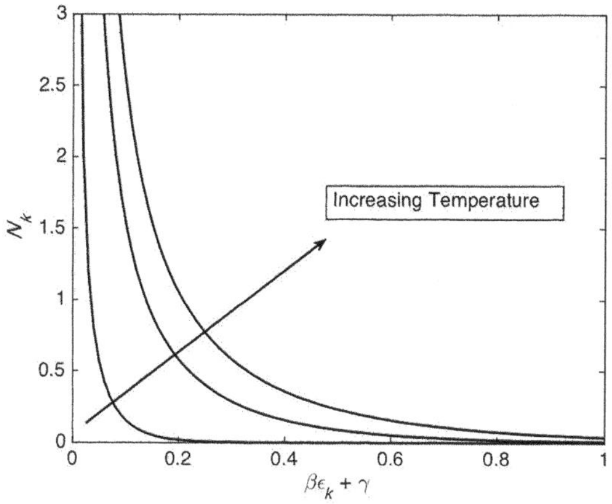

Figure 3.4 Expected value of $< N_k >$ for a Bose–Einstein system.

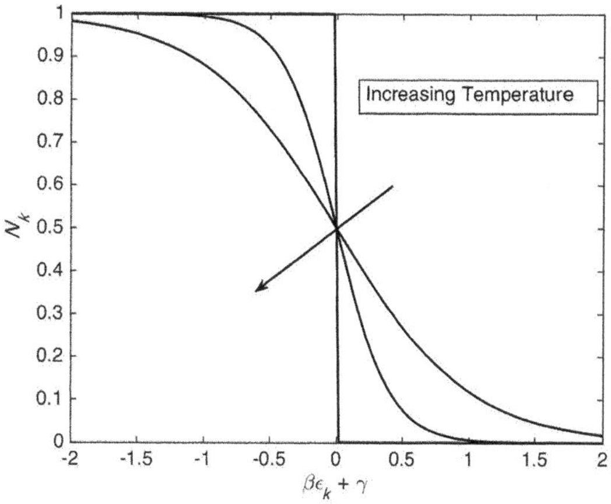

Figure 3.5 Expected value of $< N_k >$ for a Fermi–Dirac system.

Note that first, for a Bose–Einstein system, $\langle N_k \rangle$ is limited to positive values by the physical restriction that $\langle N_k \rangle$ be equal to or greater than zero. Thus, the chemical potential is limited to values less than ϵ_k. Given that, for a fixed temperature, the number of particles with a given quantum energy increases with decreasing energy. As the temperature increases, the population increases. If N is nonzero and $T = 0$ K, the chemical potential must equal ϵ_k, otherwise the population would be zero.

The behavior of a Fermi–Dirac system is substantially different from that of a Bose–Einstein system. In this case, at a fixed temperature, if $\mu \to -\infty$, then the number density of particles in the systems goes to zero, whereas if $\mu \to \infty$, then the number of particles approaches infinity. Note that the energy level takes on a special physical significance. The expectation value for N_k is always one-half, regardless of the temperature. As $T \to 0$ K, one particle occupies each quantum state with $\epsilon_k < \mu$, with no particles in higher-energy quantum states. We shall see that this behavior drives the behavior of electrons in metals and semiconductors.

3.7 Systems with Non-negligible Inter-particle Forces

If inter-particle forces become important, then a complete specification of the quantum state includes specifying the positions, or configuration, of particles. The dilemma we face when forced to consider configuration is that there is no longer a simple way to calculate the partition function.

For simplicity, consider the canonical partition function

$$Q = \sum_j e^{-\beta U_j} \tag{3.76}$$

For a system in which particle interactions are important, the energy of the member in quantum state j becomes

$$U_j = \sum_i \sum_k \epsilon_{ik} N_{ikj} + \phi(\vec{r}_1, \vec{r}_2, \vec{r}_3, \ldots, \vec{r}_N) \tag{3.77}$$

where ϕ is the energy associated with interactions between particles. Therefore,

$$Q = \sum e^{-\beta \left\{ \sum_i \sum_k \epsilon_{ik} N_{ikj} + \phi(\vec{r}_1, \vec{r}_2, \vec{r}_3, \ldots, \vec{r}_N \right\}} \tag{3.78}$$

We will show that

$$Q = \frac{q_{int} q_{tr}}{N!} Z_N \tag{3.79}$$

where q_{int} and q_{tr} are partition functions for internal motion and translational motion, respectively, and

$$Z_N = \int \ldots \int e^{-\phi/kT} d\vec{r}_1 d\vec{r}_2 \ldots d\vec{r}_N \tag{3.80}$$

is called the configuration integral. We will have to evaluate Z_N to calculate the partition function.

3.8 Summary

3.8.1 Statistics in Thermodynamics and Ensembles

This section can be summarized as: "there are lots of particles!" Therefore, we characterize the macroscopic state of a very large system of particles with probability distributions. For example, the average energy can be written as

$$\langle U \rangle = \frac{1}{n} \sum_j n_j U_j \tag{3.81}$$

where the sum is over the energy of all the members of an "ensemble," conceptually illustrated in Fig. 3.1, and n_j is the distribution of the values of the energy over the ensemble members.

3.8.2 The Postulates of Microscopic Thermodynamics

How we calculate the appropriate distribution functions is prescribed by the following postulates:

I. The macroscopic value of a mechanical–thermodynamic property of a system in equilibrium is characterized by the expected value computed from the most probable distribution of ensemble members among the allowed ensemble member quantum states.

II. The most probable distribution of ensemble members among ensemble member quantum states is that distribution for which the number of possible quantum states of the ensemble and reservoir is a maximum.

Postulate I states that we can calculate $\langle A\rangle$ using

$$\langle A\rangle = \frac{1}{n}\sum_j n_j^* A_j = \sum_j p_j A_j \tag{3.82}$$

where $p_j = n_j^*/n$ is the normalized probability distribution and * indicates that we are using the most probable distribution of n_j's or p_j's.

Postulate II describes how to determine the most probable distribution. To apply the postulate we must obtain an expression for the number of allowed quantum states as a function of the n_j and then find the n_j that maximizes this function:

$$\ln \Omega_{TOT} = \ln \Omega_R + n \ln n - \sum_j n_j \ln n_j \tag{3.83}$$

Using Lagrange's method of undetermined multipliers, we showed how to obtain the probability distribution functions.

3.8.3 The Partition Function

In this chapter we discuss at some length the type of ensemble to use. Table 3.1 shows the possibilities. It really doesn't matter which one uses, and the choice is usually made on the basis of convenience. In our case we used the grand canonical form. After the math we found that the desired probability distribution function is

$$p_j = \frac{n_j^*}{n} = \frac{e^{-\beta U_j - \sum\limits_i \gamma_i N_{ij}}}{\sum\limits_j e^{-\beta U_j - \sum\limits_i \gamma_i N_{ij}}} \tag{3.84}$$

The denominator of this expression is called the partition function, the grand partition function in this case:

$$Q_G(\beta, \gamma_i) \equiv \sum_j e^{-\beta U_j - \sum\limits_i \gamma_i N_{ij}} \tag{3.85}$$

We then showed that

$$\beta = \frac{1}{kT} \text{ and } \gamma_i = \frac{-\mu_i}{kT} \tag{3.86}$$

Were we to derive the partition functions in other representations, they would be of the form shown in Table 3.2.

3.8.4 Relationship of Partition Function to Fundamental Relation

Most importantly, we showed that the partition function is directly proportional to the fundamental relation. This takes on the general form for Massieu functions:

$$S\left[\frac{1}{T}\right] = k \ln Q \tag{3.87}$$

3.8.5 Fluctuations

Once we have the distribution function, we can calculate its statistical properties. Of most interest is the variance, or square of the standard deviation. For internal energy the variance is

$$\sigma^2(U) = \frac{1}{kT^2}\left(\frac{\partial^2 U}{\partial T^2}\right)_{V,N} \tag{3.88}$$

which for an ideal gas becomes

$$\sigma^2(U) = c_v NkT^2 \tag{3.89}$$

3.8.6 Systems with Negligible Inter-particle Forces

An important distinction in studying the macroscopic behavior of matter is the degree to which individual atoms and molecules influence their neighbors due to electrostatic interactions. If, on average, they are too far apart to influence each other, then we can neglect the effect of inter-particle forces. If that is the case, we showed that the partition function becomes

$$Q = \prod_i \prod_k \sum_{\eta=0}^{\text{Max } N_{ikj}} e^{-(\beta\varepsilon_{ik}+\gamma_i)\eta} \tag{3.90}$$

By taking into account Pauli's exclusion principle (which we discuss in more detail in the next chapter), we showed that the final form is

$$\ln Q_{FD,BE} = \sum_i \sum_k \ln\left(1 \pm e^{-\beta\varepsilon_{ik}-\gamma_i}\right)^{\pm 1} \tag{3.91}$$

where FD stands for Fermi–Dirac, particles with anti-symmetric wave functions, and BE for Bose–Einstein, particles with symmetric wave functions. These particles are called fermions and bosons, respectively, as shown below:

Symmetry	Max N_{ikj}	Particle name	System name	Examples
Anti-symmetric	1	fermion	Fermi–Dirac	D, He^3, e
Symmetric	infinity	boson	Bose–Einstein	H, He^4, photons

The most important difference between fermions and bosons is that, in a macroscopic system of fermions, only one particle is allowed to exist in any given quantum state, while for bosons there is no limit. For example, electrons are fermions, and this explains why the periodic table is organized as it is. More on this in the next chapter.

3.8.7 Systems with Non-negligible Inter-particle Forces

For systems in which interatomic forces are important, the local arrangement of particles becomes important. We call this the configuration. We will show that the partition function becomes

$$Q = \frac{q_{int} q_{tr}}{N!} Z_N \tag{3.92}$$

where q_{int} and q_{tr} are partition functions for internal motion and translational motion, respectively, and

$$Z_N = \int \dots \int e^{-\phi/kT} d\vec{r}_1 d\vec{r}_2 \dots d\vec{r}_N \tag{3.93}$$

is called the configuration integral. We will have to evaluate Z_N to calculate the partition function.

3.9 Problems

3.1 Consider four hands of cards, each hand containing 13 cards with the 52 cards forming a conventional deck. How many different deals are possible? Simplify your result using Stirling's approximation. The order of the cards in each hand is not important.

3.2 Calculate and plot the error in using Stirling's approximation $\ln x! \simeq x \ln x - x$ for $1 < x < 100$.

3.3 Use the method of Langrangian multipliers to maximize $-\sum_i p_i \ln p_i$ subject to the constraint $\sum_i p_i = 1$. Show that when this quantity is a maximum, $p_i =$ constant.

3.4 Using Lagrange's method of undetermined multipliers, show that the partition function for a canonical ensemble is

$$Q = \sum_j \exp\left(-\frac{U_j}{kT}\right)$$

3.5 Show that for a system of indistinguishable particles (single component) with negligible inter-particle forces,

$$S = k[\langle N_k \rangle \ln \langle N_k \rangle \pm (1 + \langle N_k \rangle) \ln(1 \mp \langle N_k \rangle)]$$

Hint: Start with the Massieu function for the grand canonical representation, then solve for S and substitute in the quantum statistical expressions for U and N.

3.6 Prove that for a system of indistinguishable particles (single component) with negligible inter-particle forces,

$$\ln Q_{FD} < \ln Q_{MB} < \ln Q_{BE}$$

3.7 Show that Eqs (3.66) and (3.67) are equivalent.

3.8 If we wished to do so, we could take $\overline{\Delta U^3} \equiv \overline{(Uj - \bar{U})^3}$ as a measure of energy fluctuations in the canonical ensemble. Show that

$$\overline{\Delta U^3} = -\frac{\partial^3 \ln Q}{\partial \beta^3}$$

3.9 At what approximate altitude would a 1-cm^3 sample of air exhibit 1% fluctuations in internal energy? Use data from the US Standard Atmosphere (see engineeringtoolbox.com).

4 Quantum Mechanics

In this chapter we learn enough about quantum mechanics to gain a conceptual understanding of the structure of atoms and molecules and provide quantitative relationships that will allow us to compute macroscopic properties from first principles. Quantum mechanics developed as a consequence of the inability of classical Newtonian mechanics and Maxwell's electromagnetic theory to explain certain experimental observations. The development of quantum mechanics is a fascinating episode in the history of modern science. Because it took place over a relatively short period from the end of the nineteenth century through the first third of the twentieth century, it is well documented and is the subject of many books and treatises. See, for example, Gamow [24]. In the next section we give an abbreviated version of the history, outlining the major conceptual hurdles and advances that led to our modern understanding. We then introduce four postulates that describe how to calculate quantum-mechanical properties. The remainder of the chapter discusses specific atomic and molecular behaviors. (This chapter largely follows Incropera [7].)

4.1 A Brief History

We start with a brief history of the atom. The Greek philosopher Democratus [25] was the first to develop the idea of atoms. He postulated that if you divided matter over and over again, you would eventually find the smallest piece, the atom. However, his ideas were not explored seriously for more than 2000 years. In the 1800s scientists began to carefully explore the behavior of matter. An English scientist, John Dalton, carried out experiments that seemed to indicate that matter was composed of some type of elementary particles [26]. In 1897 the English physicist J. J. Thomson discovered the electron and proposed a model for the structure of the atom that included positive and negative charges. He recognized that electrons carried a negative charge but that matter was neutral. (His general ideas were confirmed in a series of scattering experiments over the period 1911–1919 by Rutherford, which firmly established the presence of electrons and protons.)

In 1861–1862 James Clerk Maxwell [18] published early forms of what are now called "Maxwell's equations." These equations describe how electric and magnetic fields interact with charged matter and how electromagnetic radiation is generated. Electromagnetic radiation, which includes visible light, was described as a wave phenomenon in which oscillating electric and magnetic fields interact. It is considered a "classical" theory in this sense. For an electromagnetic wave, there is a fixed relation between the frequency and the wavelength given by

$$\nu = \frac{c}{\lambda} \tag{4.1}$$

where ν is the frequency, λ is the wavelength, and c is the speed of light.

By the end of the nineteenth century, then, it was recognized that matter was composed of elementary particles, or atoms, that contained positive and negative charges, and that light (more generally radiation) was composed of electromagnetic waves. Both were viewed through the lens of classical theory.

4.1.1 Wave–Particle Duality – Electromagnetic Radiation Behaves Like Particles

4.1.1.1 Blackbody Radiation

The first experiment to challenge classical theories was the observation of the spectrum of blackbody radiation. It was known that any surface with a temperature greater than absolute zero emits radiation. A black body is one that absorbs all radiation striking it. Such a surface can be approximated by a cavity with highly absorbing internal walls. If a small hole is placed in the wall, any radiation entering will reflect numerous times before exiting, and will be absorbed almost entirely. At the same time, the atoms making up the walls are vibrating. Classical electromagnetic theory predicts that when two charges oscillate with respect to each other, they emit radiation. At equilibrium, the amount of radiation energy entering the cavity must equal the amount leaving. Assuming the cavity walls are at a fixed temperature T, the radiation leaving the cavity is referred to as blackbody emission.

The spectral energy distribution of blackbody radiation is shown in Fig. 4.1. As can be seen, the energy per unit wavelength approaches zero at high frequencies, reaches a peak, and then declines again at low frequencies. Every attempt to predict this functional relationship using classical theory failed. The curve labeled "Rayleigh–Jeans formula" is based on classical electromagnetic theory. It matched the long-wavelength region of the spectrum but diverged sharply as the wavelength decreased. At the time, this was called the "ultraviolet catastrophe." Wien's formula, on the other hand, predicted the high-frequency behavior well, but failed at lower frequencies.

In 1901, Max Planck provided the first important concept that led to modern physics. He did what any engineer would do; he performed an empirical curve fit to the data. The correlation was of the form

$$u_\nu = \frac{8\pi h\nu^3}{c^3}\frac{1}{(e^{-h\nu/kT} - 1)} \tag{4.2}$$

The constant h is now known as Planck's constant and is equal to 6.626×10^{-34} J-sec. After some thought, he concluded that the energy associated with the atomic oscillators could not be continuously distributed. This became known as "Planck's postulate," the essence of which is that

$$\epsilon_n = nh\nu \tag{4.3}$$

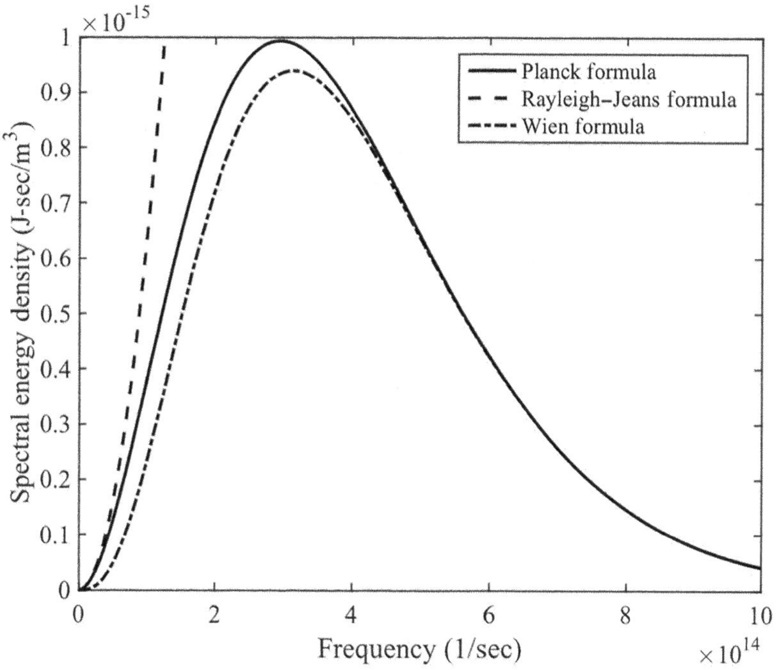

Figure 4.1 Spectral distribution of blackbody radiation.

where n is an integer ($n = 1, 2, 3, \ldots$). Thus, the oscillator energy is "quantized." If the energy of an oscillator is quantized, then it may only exchange energy in discrete amounts.

4.1.1.2 The Photoelectric Effect

Consider the experimental arrangement illustrated in Fig. 4.2. The experiment consists of illuminating a metallic surface with radiation of fixed frequency and measuring the current due to electrons that are emitted from the surface because of radiation/surface interactions. What was observed is that below a certain frequency no electrons were emitted, regardless of the intensity of the radiation. Classical theory would predict that as the intensity of radiation is increased, the oscillators would acquire more and more energy, eventually gaining enough kinetic energy that an electron would escape the surface. Einstein, in 1905, proposed an alternative theory in which the incident radiation behaved not as a wave, but rather as particles. He called these particles "photons" or "quanta" of light. The energy of each photon is

$$\epsilon = h\nu \tag{4.4}$$

where h is Planck's constant. The energy of an electron emitted from the surface is then (T is historically used to denote the electron energy)

$$T = h\nu - V - W \tag{4.5}$$

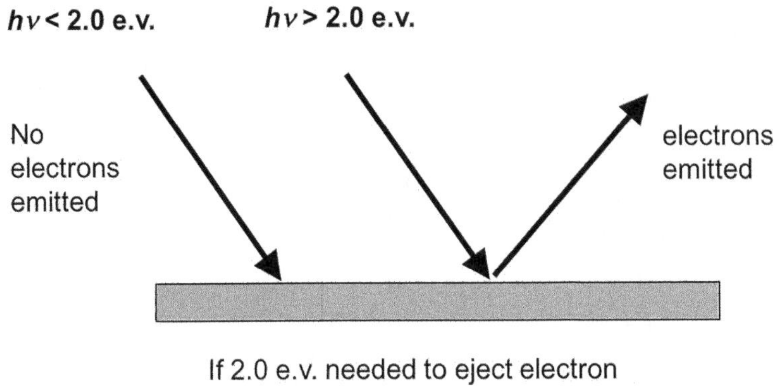

Figure 4.2 Photoelectric emission.

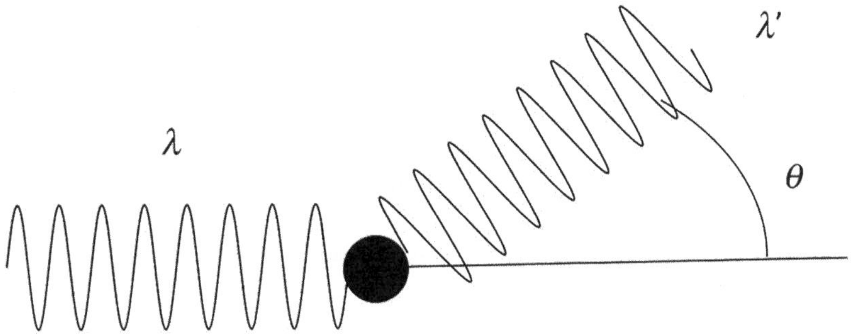

Figure 4.3 The Compton effect.

where V is the energy required to bring the electron to the surface and W is the energy to eject it from the surface. W is called the work function and depends on the material properties of the surface. Therefore, unless the photon energy is large enough to overcome the work function and V, no electron emission will take place.

Einstein's theory completely explained the photoelectric effect. However, the scientific community was then faced with the so-called "wave–particle" duality of radiation.

4.1.1.3 The Compton Effect

In 1923, Arthur Compton published the results of an experiment involving the scattering of X-rays by a metallic foil (Fig. 4.3). It was observed that for X-rays of a given wavelength, the scattered X-rays emerged at a difference wavelength that depended on the scattering angle θ. Compton proposed that the "photons" scattered like "particles." He derived the relation

$$\lambda' - \lambda = \frac{h}{m_e c}(1 - \cos\theta) \tag{4.6}$$

where λ is the incident wavelength, λ' the scattered wavelength, and m_e the mass of an electron. The theory was based on photons carrying the property of momentum in the amount

$$\epsilon = h/\lambda \tag{4.7}$$

Compton's finding removed any doubt regarding the dual nature of radiation.

4.1.2 Particle–Wave Duality – Particles Can Display Wave-Like Behavior

4.1.2.1 The De Broglie Postulate

In 1924 Louis de Broglie postulated that particles can behave like waves. Waves typically have the properties of wavelength, phase-velocity, and group-velocity. Properties associated with matter before quantum mechanics were mass, velocity, and momentum. A connection between these two worlds is the de Broglie relation

$$\lambda = h/p \tag{4.8}$$

which attributes a wavelength to a particle with momentum p. This can also be written as

$$p = \hbar k \tag{4.9}$$

where $\hbar = h/2\pi$ and $k = 2\pi/\lambda$ is the wavenumber. The kinetic energy of a classical particle can then be written as a function of the wavenumber

$$K.E. = \frac{p^2}{2m} = \frac{\hbar^2 k^2}{2m} \tag{4.10}$$

4.1.2.2 Davisson–Germer Experiment

In 1927 Davisson and Germer confirmed the wave-like nature of matter. They observed that a beam of electrons with momentum p was scattered by a nickel crystal like X-rays of the same wavelength. The relation between momentum and wavelength was given by the de Broglie relation [Eq. (4.8)].

4.1.3 Heisenberg Uncertainty Principle

In 1927, Heisenberg developed what has come to be called the "Heisenberg uncertainty principle." It is based on a wave-packet representation of a particle. A wave-packet is a wave-like structure that is spatially localized, as illustrated in Fig. 4.4.

A so-called "plane wave" can be described as

$$\psi(x) \propto e^{ik_0x} = e^{ip_0x/\hbar} \tag{4.11}$$

where we have used the de Broglie relation to relate wavenumber and momentum. A plane wave, however, has infinite extent, but a wave similar to that shown in Fig. 4.4 can be represented by an expansion of plane waves:

$$\psi(x) \propto \sum_n A_n e^{ip_nx/\hbar} \tag{4.12}$$

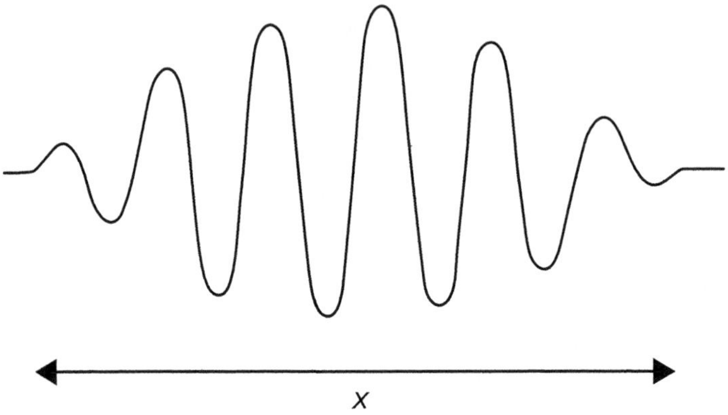

Figure 4.4 A wave packet.

where the A_n are expansion coefficients that represent the relative contribution of each mode p_n to the overall wave. In the continuum limit of an infinite number of modes, one can define a wave function

$$\psi(x) = \frac{1}{\sqrt{2\pi\hbar}} \int_{-\infty}^{\infty} \phi(p) e^{ipx/\hbar} dp \tag{4.13}$$

This form of the wave function is normalized, that is the integral is unity. Since the probability of finding the particle between x locations a and b is

$$P[a \leq x \leq b] = \int_{a}^{b} |\psi(x)|^2 dx \tag{4.14}$$

the normalization insures that the particle is somewhere in space. However, for a wave packet with finite width, this integral will have a value only over the spatial domain of the particle. Its average position will be

$$\bar{x} = \int_{-\infty}^{\infty} x|\psi(x)|^2 dx \tag{4.15}$$

and the dispersion is

$$D_x = \int_{-\infty}^{\infty} x^2 |\psi(x)|^2 dx \tag{4.16}$$

It turns out that the function $\phi(p)$ is the Fourier transform of $\psi(x)$. Were we to calculate the dispersion in the momentum, we would obtain the result that

$$\Delta x \Delta p \geq \frac{\hbar}{2} \tag{4.17}$$

This is Heisenberg's uncertainty principle. It states that the position and momentum of a particle cannot be measured simultaneously with arbitrarily high precision; there is a minimum uncertainty in the product of the two. It comes about purely from the wave character of the particle.

4.2 The Postulates of Quantum Mechanics

By the early 1930s, the foundations of modern quantum mechanics were fairly well established. It then became reasonable to consolidate the theory with a set of postulates. (Here, system means a single particle or a collection of particles such as an atom or molecule.)

I. **Each system can be characterized by a wave function, which may be complex**

$$\Psi(\vec{r}, t) \tag{4.18}$$

and which contains all the information that is known about the system.

II. **The wave function is defined such that**

$$p(\vec{r}, t)dV = \Psi^*(\vec{r}, t)\Psi(\vec{r}, t)dV = |\Psi(\vec{r}, t)|^2 \tag{4.19}$$

is the probability of finding the system in the volume element dV. The consequence of this is that

$$\int_{-\infty}^{\infty} \Psi^*(\vec{r}, t)\Psi(\vec{r}, t)dV = 1 \tag{4.20}$$

That is, the system will be found somewhere.

III. **With every dynamical variable, there are associated operators**

$$\vec{r}\colon \vec{r}_{op} = \vec{r} \tag{4.21}$$

$$\vec{p}\colon \vec{p}_{op} = \vec{p} \tag{4.22}$$

$$\epsilon\colon \epsilon_{op} = \epsilon \tag{4.23}$$

$$B(\vec{r}, t)\colon B_{op} = B(\vec{r}, -i\hbar\vec{\nabla}) \tag{4.24}$$

IV. **The expectation value of any physical observable is**

$$\langle B\rangle = \int_{-\infty}^{\infty} \Psi^*(\vec{r}, t)B_{op}\Psi(\vec{r}, t)dV \tag{4.25}$$

The wave function is described by Schrödinger's equation, which starts with the classical form for conservation of energy (H is the Hamiltonian of classical mechanics)

$$H = \frac{p^2}{2m} + V(\vec{r}, t) = \epsilon \tag{4.26}$$

and substitutes the operators so that

$$-\frac{\hbar^2}{2m}\nabla^2\Psi(\vec{r}, t) + V(\vec{r}, t)\Psi(\vec{r}, t) = i\hbar\frac{\partial\Psi(\vec{r}, t)}{\partial t} \tag{4.27}$$

Schrödinger chose this form for two reasons:

1. He needed a linear equation to describe the propagation of wave packets.
2. He wanted the equation to be symmetric with the classical conservation of energy statement.

4.3 Solutions of the Wave Equation

As thermodynamicists, we are interested in obtaining the allowed quantum states of individual atoms and molecules so that we may calculate the allowed quantum states of ensemble members. The complete specification of the atomic or molecular quantum state requires understanding translational and electronic motion in the case of atoms, with rotational and vibration motion also important for molecules.

The wave equation is a partial differential equation, first order in time and second order in space. The potential energy function acts as a source term, and can take on a variety of forms depending on the circumstances. For isolated atoms and molecules, the potential energy will be either zero, or a function of the relative positions of subatomic or molecular particles, but not time. As we shall see, this leads to the ability to solve the wave equation using the method of separation of variables, familiar from solving the heat equation. The solutions in this case lead to stationary (constant) expectation values for the mechanical, dynamical variables. In real macroscopic systems, the atoms and molecules interact with each other at very fast rates, via collisions in the gas phase, or through close, continuous contact in the liquid and solid phases. However, between collisions, they rapidly relax to stationary states, and the properties of those states dominate the overall thermodynamic behavior. As a result, we will focus our attention on stationary solutions that describe translational, rotational, vibrational, and electronic behavior.

We start, then, by assuming that the potential energy term V is a function only of position:

$$V(\vec{r},t) = V(\vec{r}) \tag{4.28}$$

As a result, the wave equation can be written as

$$i\hbar\frac{\partial \Psi(\vec{r},t)}{\partial t} = -\frac{\hbar^2}{2m}\nabla^2\Psi(\vec{r},t) + V(\vec{r})\Psi(\vec{r},t) \tag{4.29}$$

We assume that the wave function can be written as a product of a time-dependent function and a spatially dependent function:

$$\Psi(\vec{r},t) = \phi(t)\psi(\vec{r}) \tag{4.30}$$

Substituting this into the wave equation and then dividing by the same function, we obtain

$$\frac{i\hbar}{\phi}\frac{\partial \phi}{\partial t} = \frac{1}{\psi}\left[-\frac{\hbar^2}{2m}\nabla^2\psi + V(\vec{r})\psi\right] \tag{4.31}$$

Since the two sides are functions of different variables, they must be equal to within an arbitrary constant, so that the time-dependent wave equation becomes

$$\frac{i\hbar}{\phi}\frac{\partial \phi}{\partial t} = C \tag{4.32}$$

and the spatially dependent wave equation becomes

$$\frac{1}{\psi}\left[-\frac{\hbar^2}{2m}\nabla^2\psi + V(\vec{r})\psi\right] = C \tag{4.33}$$

The time-dependent part is a simple first-order, ordinary differential equation, and can be solved once and for all. Its solution is

$$\phi(t) = \exp\left(-i\frac{C}{\hbar}t\right) \tag{4.34}$$

Therefore,

$$\Psi(\vec{r},t) = \psi(\vec{r})\exp\left(-i\frac{C}{\hbar}t\right) \tag{4.35}$$

It turns out that the separation constant C has a very important physical meaning which we can understand if we use the wave function to calculate the expectation value of the energy:

$$\langle\epsilon\rangle = \int_V \Psi^*\left(i\hbar\frac{\partial}{\partial t}\right)\Psi dV = \int_V \psi^* e^{i\frac{C}{\hbar}t} C\psi e^{-i\frac{C}{\hbar}t} dV = C \tag{4.36}$$

Thus, C is equal to the expectation value of the energy, and the spatially dependent wave equation becomes

$$-\frac{\hbar^2}{2m}\nabla^2\psi + V(\vec{r})\psi = \epsilon\psi \tag{4.37}$$

This equation is also called the stationary wave equation, because its solutions result in expectation values of the dynamical properties that do not depend on time. To solve it requires two boundary conditions in space, and knowledge of the potential energy function.

We can solve for the motion of a single particle using the above equations. For a two-particle system, say a diatomic molecule or an atom with a single electron, the energy becomes

$$\epsilon = \frac{p_1^2}{2m_1} + \frac{p_2^2}{2m^2} + V(\vec{r}) \tag{4.38}$$

One can show [7] that transforming the wave equation into center of mass coordinates results in the following equations for external and internal motion:

$$\frac{\hbar^2}{2m_t}\nabla^2\psi_e + \epsilon_e\psi_e = 0 \tag{4.39}$$

$$\frac{\hbar^2}{2\mu}\nabla^2\psi_{int} + (\epsilon_e - V(\vec{r}))\psi_e = 0 \tag{4.40}$$

where

$$\mu = \frac{m_1 m_2}{m_1 + m_2} \tag{4.41}$$

is the reduced mass.

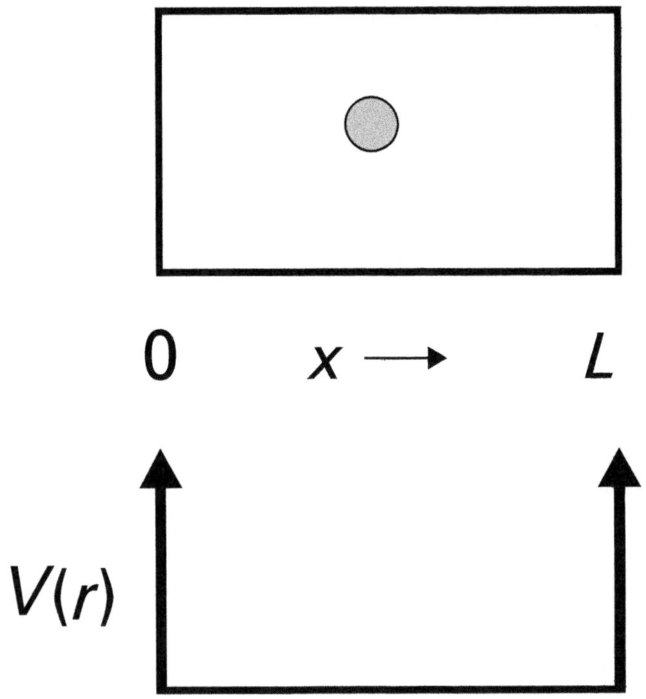

Figure 4.5 The particle in a box.

4.3.1 The Particle in a Box

We first consider the linear motion of a single particle in a square box, as illustrated in Fig. 4.5. While we expect that $V(x)$ is zero within the box, we also expect that the particle will remain in the box. Therefore, $V(x)$ cannot be zero at the walls. Indeed, $V(x)$ must be essentially infinite at the walls, so that the particle experiences a strong opposing force as it approaches the wall. This is illustrated conceptually in the lower half of Fig. 4.5, where $V(x)$ is represented as delta functions at the walls.

The wave equation in one dimension becomes

$$\frac{\hbar^2}{2m}\frac{\partial \psi^2}{\partial^2 x} = -\epsilon_x \psi \tag{4.42}$$

The boundary conditions are quite simple; the wave function must be zero at each boundary. This can be deduced from the second postulate, which states that the probability of finding the particle within a given differential volume is

$$p(\vec{r},t)dV = \Psi^*(\vec{r},t)\Psi(\vec{r},t)dV = \psi^*(\vec{r})\psi(\vec{r})dV \tag{4.43}$$

As a result, we have

$$\psi(0) = \psi(L) = 0 \tag{4.44}$$

The solution to Eq. (4.42) is simply

$$\psi(x) = A \sin\left[\sqrt{\frac{2m\epsilon_x}{\hbar^2}}x\right] + B\cos\left[\sqrt{\frac{2m\epsilon_x}{\hbar^2}}x\right] \tag{4.45}$$

Note that the first condition requires that B be zero, and the second that

$$0 = A\sin\left[\sqrt{\frac{2m\epsilon_x}{\hbar^2}}L\right] \tag{4.46}$$

For this relation to be satisfied, the argument of the sin must be equal to zero or a multiple of π. The only parameter within the argument that is not a fixed constant of the problem is the energy ϵ_x. Therefore, while it is physically restricted to being zero or positive, it must also be restricted to values that meet the boundary conditions. These values become

$$\epsilon_x = \frac{\hbar^2\pi^2}{2mL}n_x^2 \tag{4.47}$$

where n_x is zero or a positive integer. It is because not all energies are allowed that we use the term "quantum." The constant A is evaluated by normalizing the wave function

$$\int_0^L |\psi|^2 dx = 1 \tag{4.48}$$

so that

$$A = \sqrt{\frac{2}{L}} \tag{4.49}$$

and the stationary wave function becomes

$$\psi(x) = \sqrt{\frac{2}{L}}\sin\left(\frac{\pi n_x x}{L}\right), \quad n_x = 0, 1, 2, \ldots \tag{4.50}$$

The three-dimensional problem is now easily solved. Because the potential function is zero, the three coordinate directions are independent of each other. Therefore, the solutions for the y and z directions are the same as for the x direction, assuming the appropriate dimension is used. Assuming that the box is a cube, we thus have

$$\epsilon = \epsilon_x + \epsilon_y + \epsilon_z = \frac{h^2}{8mV^{2/3}}(n_x^2 + n_y^2 + n_z^2) \tag{4.51}$$

and

$$\psi(x) = \sqrt{\frac{8}{L^3}}\sin\left(\frac{\pi n_x x}{L}\right)\sin\left(\frac{\pi n_y y}{L}\right)\sin\left(\frac{\pi n_z z}{L}\right) \tag{4.52}$$

An important finding is that the energy is a function of the volume of the box. Indeed, we will see that for an ideal gas, this leads to all the volume dependency in the fundamental relation.

It is very important to note that different combinations of the three quantum numbers can result in the same energy. This is illustrated in Table 4.1. When more than one

Table 4.1 Degeneracy of translational quantum states

n_x	n_y	n_z	$n_x^2+n_y^2+n_z^2$	g
1	1	1	3	1
2	1	1	6	
1	2	1	6	3
1	1	2	6	

quantum state have the same energy, they are often lumped together and identified as an "energy state" in contrast to a "quantum state" or "eigenstate." The energy state is then said to be degenerate. The letter g is used to denote the value of the degeneracy. In Table 4.1, for example, the first state listed is not degenerate; however, the next three have the same total quantum number and thus energy, and the degeneracy is three. For large values of n, the degeneracy can be very large. We shall see that the average kinetic energy of an atom or molecule is equal to

$$\langle\epsilon\rangle = \frac{3}{2}kT \tag{4.53}$$

where k is Boltzmann's constant. Thus, a typical value of the translational quantum number at room temperature is about 10^8. [Calculate ϵ for 300 K and then use Eq. (4.51) to estimate an average quantum number.] We will work out what the degeneracy would be as a function of n_{trans} in Chapter 5.

4.3.2 Internal Motion

To explore the internal motions of atoms and molecules, we must take into account the fact that even the most simple models involve multiple particles. Atoms are composed of nuclei and electrons, and molecules of multiple atoms. We will only explore analytic solutions for the simplest cases, the one-electron or hydrogen atom and the diatomic molecule. We will resort to numerical methods to explore more realistic behavior.

We start with the stationary wave equation

$$\frac{\hbar^2}{2\mu}\nabla^2\psi_{int} + (\epsilon_{int} - V(\vec{r}))\psi_{int} = 0 \tag{4.54}$$

Converting this equation into spherical coordinates (Fig. 4.6), we obtain

$$\begin{aligned}\left[\frac{1}{r^2}\frac{\partial}{\partial r}\left(r^2\frac{\partial}{\partial r}\right) + \frac{1}{r^2\sin^2\theta}\frac{\partial^2}{\partial\phi^2} + \frac{1}{r^2\sin\theta}\frac{\partial}{\partial\theta}\left(\sin\theta\frac{\partial}{\partial\theta}\right)\right]\psi_{i\,int}&\\ + \frac{2\mu}{\hbar^2}\left[\epsilon_{int} - V_{int}(r)\right]\psi_{int} = 0&\end{aligned} \tag{4.55}$$

Again we pursue separation of variables, defining

$$\psi_{int}(r,\theta,\phi) = R(r)Y(\theta,\phi) \tag{4.56}$$

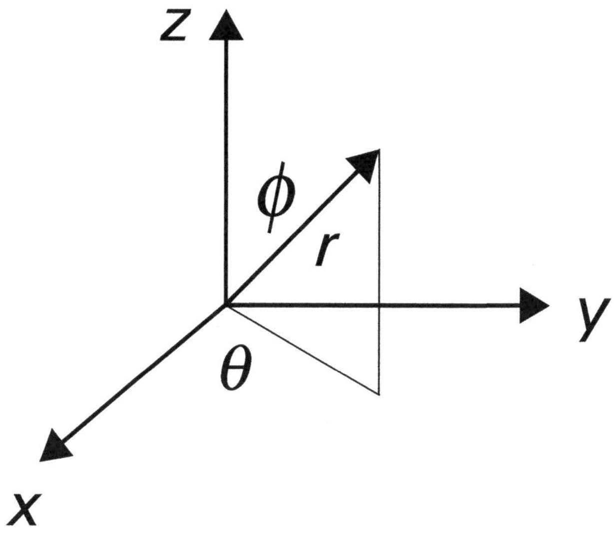

Figure 4.6 Spherical coordinate system.

and inserting into Eq. (4.55):

$$\frac{1}{R}\frac{d}{dr}\left(r^2\frac{dR}{dr}\right)+\frac{2\mu r^2}{\hbar^2}\left[\epsilon_{int}-V_{int}(r)\right]=-\frac{1}{Y}\left[\frac{1}{\sin\theta}\frac{\partial}{\partial\theta}\left(\sin\theta\frac{\partial Y}{\partial\theta}\right)+\frac{1}{\sin^2\theta}\frac{\partial^2 Y}{\partial\phi^2}\right] \tag{4.57}$$

Calling the separation variable α, we get a radial equation and one that contains all the angular dependence:

$$\frac{d}{dr}\left(r^2\frac{dR}{dr}\right)+\left\{\frac{2\mu r^2}{\hbar^2}\left[\epsilon_{int}-V_{int}(r)\right]-\alpha\right\}R=0 \tag{4.58}$$

$$\left[\frac{1}{\sin\theta}\frac{\partial}{\partial\theta}\left(\sin\theta\frac{\partial Y}{\partial\theta}\right)+\frac{1}{\sin^2\theta}\frac{\partial^2 Y}{\partial\phi^2}\right]=-\alpha Y \tag{4.59}$$

Now, separate the angular wave equation using

$$Y(\theta,\phi)=\Theta(\theta)\Phi(\phi) \tag{4.60}$$

resulting in

$$\frac{d^2\Phi}{d\phi^2}+\beta\Phi=0 \tag{4.61}$$

and

$$\frac{1}{\sin\theta}\frac{d}{d\theta}\left(\sin\theta\frac{\partial\Theta}{\partial\theta}\right)+\left(\alpha-\frac{\beta}{\sin^2\theta}\right)\Theta=0 \tag{4.62}$$

where β is the separation constant. Note that neither equation depends on the potential energy term, $V(r)$, so they can be solved once and for all. The general solution for Φ is simply

$$\Phi(\phi) = \exp(i\beta^{1/2}\phi) \tag{4.63}$$

Note that this function must be continuous and single valued, so

$$\Phi(\phi) = \Phi(\phi + 2\pi) \tag{4.64}$$

This will only be true if $\beta^{1/2}$ is equal to a constant, which we will call m_l, where $m_l = 0, \pm 1, \pm 2, \ldots$ Therefore,

$$\Phi(\phi) = \exp(im_l\phi) \tag{4.65}$$

Plugging the expression for β into the equation for Θ:

$$\frac{1}{\sin\theta}\frac{\partial}{\partial\theta}\left(\sin\theta\frac{\partial\Theta}{\partial\theta}\right) + \left(\alpha - \frac{m_l^2}{\sin^2\theta}\right)\Theta = 0 \tag{4.66}$$

This is a Legendre equation. (Legendre was a nineteenth-century mathematician who worked out the series solution of equations of this form.) To satisfy the boundary conditions we must have $\alpha = l(l+1)$, $l = i + |m_l|$, $i = 0, 1, 2, 3, \ldots$ The solution is the associated Legendre function

$$\Theta = P_l^{|m_l|}(\cos\theta) = \frac{1}{2^l l!}(1 - \cos^2\theta)^{|m_l|/2}\frac{d^{|m_l|+l}}{d(\sin\theta)^{|m_l|+l}}(\sin^2\theta - 1)^l \tag{4.67}$$

Thus, the angular component of the wave equation becomes

$$Y(\theta, \phi) = C_{l,m_l}P_l^{|m_l|}(\cos\theta)e^{im_l\phi} \tag{4.68}$$

where C_{l,m_l} is a normalization constant:

$$C_{l,m_l} = \frac{1}{(2\pi)^{1/2}}\left[\frac{(2l+1)(l-|m_l|)!}{2(l+|m_l|)!}\right]^{1/2} \tag{4.69}$$

The integer constants l and m_l turn out to be quantum numbers associated with angular momentum. Recall Eq. (4.25):

$$\langle B\rangle = \int_{-\infty}^{\infty}\Psi^*(\vec{r}, t)B_{op}\Psi(\vec{r}, t)dV \tag{4.70}$$

If we calculated the angular momentum of our system, we would get

$$\langle L^2\rangle = l(l+1)\hbar^2 \tag{4.71}$$

Thus l determines the expectation value of the angular momentum and is called the angular momentum quantum number. Likewise, if we calculated the z component of angular momentum, we would obtain

$$\langle L_z\rangle = m_l\hbar \tag{4.72}$$

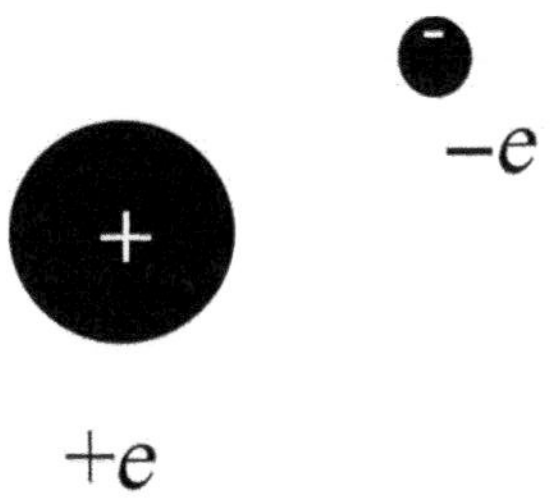

Figure 4.7 The hydrogenic atom.

Normally, atoms or molecules are randomly oriented. However, in the presence of a magnetic field they will align. Thus, m_l is called the magnetic quantum number.

4.3.3 The Hydrogenic Atom

We can obtain a relatively simple and illustrative, although inaccurate, solution to the radial wave equation of an atom if we assume that the atom is composed of a positively charged nucleus and a single electron, as illustrated in Fig. 4.7, where e is the electronic charge 1.602×10^{-19} C. We also assume that the potential energy term is given by Coulomb's law:

$$V(r) = \int_{\infty}^{r} \frac{e^2}{4\pi\epsilon_0} \frac{dr}{r^2} = -\frac{e^2}{4\pi\epsilon_0 r} \tag{4.73}$$

Starting with the radial wave equation

$$\frac{d}{dr}\left(r^2 \frac{dR}{dr}\right) + \left\{\frac{2\mu r^2}{\hbar^2}\left[\epsilon_{int} - V_{int}(r)\right] - \alpha\right\} R = 0 \tag{4.74}$$

and inserting V and α (the two-particle separation constant), we obtain

$$\frac{1}{r^2}\frac{d}{dr}\left(r^2 \frac{dR}{dr}\right) - \left\{\frac{l(l+1)}{r^2} + \frac{2\mu}{\hbar^2}\left[\epsilon_{int} + \frac{e^2}{4\pi\epsilon_0 r}\right]\right\} R = 0 \tag{4.75}$$

It is useful to transform this equation so that

$$\frac{1}{\rho^2}\frac{d}{d\rho}\left(\rho^2 \frac{dR}{d\rho}\right) - \left[\frac{l(l+1)}{\rho^2} - \frac{n}{\rho} + \frac{1}{4}\right] R = 0 \tag{4.76}$$

where $\rho = \frac{2r}{na_0}$, $n^2 = -\frac{\hbar^2}{2\mu a_0^2 \epsilon_{el}}$, and $a_0 = -\frac{\hbar^2}{\pi \mu e^2}$.

Equation (4.76) is an example of an ordinary differential equation that is amenable to solution using series methods. Many such equations were solved in the nineteenth century by mathematicians, this one by Edmond Laguerre. The solution is of the form

$$R_{nl}(\rho) = -\left\{\frac{(n-l-1)!}{2n[(n+l)!]^3}\right\}^{1/2} \left(\frac{2}{na_0}\right)^{3/2} \rho^l \exp(-\rho/2) L_{n+1}^{2l+1}(\rho) \tag{4.77}$$

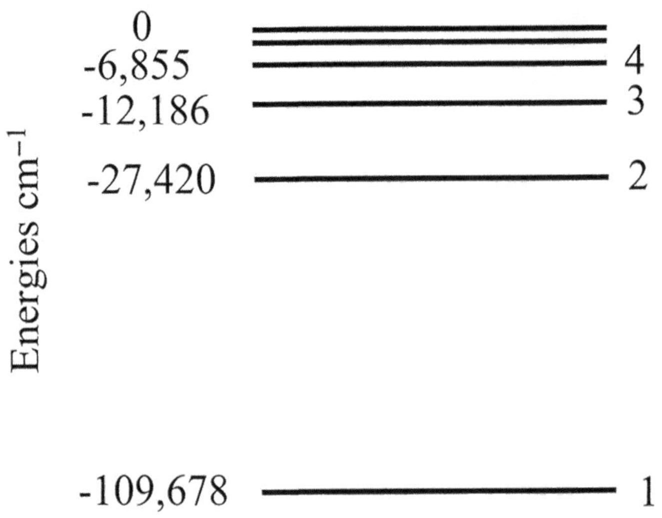

Figure 4.8 Hydrogen atom energy levels.

where

$$L_{n+1}^{2l+1}(\rho) = \sum_{k=0}^{n-l-1} (-1)^{k+1} \frac{[(n+l)!]^2}{(n-l-1-k)!\,(2l+1+k)!\,k!}\rho^k \tag{4.78}$$

is the associated Laguerre polynomial (see Abramowitz and Stegun [27]).

Again we obtain an index, n, called the principal quantum number. It can take on values $n = 1, 2, 3, \ldots$ Values of l, the angular momentum quantum number, are restricted to being less than n, or $l = 0, 1, 2, \ldots, n-1$.

Solving for the energy, we obtain (where Z is the nuclear charge)

$$\epsilon_n = -\frac{Z^2 e^4 \mu}{32\pi^2\epsilon_0^2\hbar^2}\frac{1}{n^2}\,,\; n = 1, 2, 3, \ldots \tag{4.79}$$

This result is illustrated in Fig. 4.8. Note that by convention, energy is referenced to zero for $n \to \infty$ where the nucleus and electron are infinitely far apart. (In that case we say the atom is ionized.) $n = 1$ is called the ground state and is the state with the lowest energy.

For this ideal hydrogenic atom, the other quantum numbers, l and m_l, do not affect the energy. However, their values are limited by the value of n. One can show that the allowed values of l and m_l are limited as

$$l = 0, 1, 2, 3, \ldots, n-1;\; l = i + |m_l|\,;\; i = 0, 1, 2, \ldots \tag{4.80}$$

so that

$$g_n = n^2 \tag{4.81}$$

One further complication involves the electrons. Quantum mechanics tells us that electrons, in addition to orbiting the nuclei, have spin. The spin angular momentum is given by

$$S^2 = s(s+1)\hbar^2 \tag{4.82}$$

where s is the spin angular momentum quantum number. The value of s is limited to 1/2. As with the general case,

$$S_z = m_s\hbar \tag{4.83}$$

where

$$m_s = \pm\frac{1}{2} \tag{4.84}$$

Therefore, for our simple hydrogenic atom the total degeneracy is actually

$$g_n = 2n^2 \tag{4.85}$$

4.3.4 The Born–Oppenheimer Approximation and the Diatomic Molecule

In general, the analytic solution of the wave equation for molecules is a complex, time-dependent problem. Even for a diatomic molecule it is a many-body problem, since all the electrons must be accounted for. However, there is a great simplification if the electronic motion can be decoupled from the nuclear motion, as pointed out by Max Born and J. Robert Oppenheimer in 1927. This turns out to be possible because the mass of an electron is 1/1830th that of a proton. This means that the electrons are moving much faster than the nuclei, and approach steady motion rapidly. As a result, the effect of the electrons on the nuclei is an effective force between the nuclei that only depends on the distance between them. Thus, the force can be described as due to a potential field. We know that stable potentials (i.e. ones that can result in chemical bonding) are attractive at a distance and repulsive as the nuclei approach each other. One simple function that contains this behavior is the Morse potential, which is illustrated in Fig. 4.9:

$$V(r) = D_e\left[1 - e^{-\beta(r-r_e)}\right]^2 \tag{4.86}$$

Real potentials must either be determined experimentally or from numerical solutions of the wave equation. However, most bonding potentials look more or less like Fig. 4.9. If we concern ourselves with molecules in quantum states such that only the lower-energy portion of the potential is important, then the typical potential looks much like that for a harmonic oscillator (Fig. 4.9). In this case the force between the nuclei is

$$F = k(r - r_e) \tag{4.87}$$

where r_e is the equilibrium nuclear spacing and k is a force constant. The potential energy is then

$$V(r) = \frac{k(r - r_e)^2}{2} \tag{4.88}$$

We now consider the two extra modes of motion available to a molecule. The first is rotation (Fig. 4.10). In a real molecule, rotation will exert a centripetal force that will cause the bond between the nuclei to stretch. However, if we assume that the two

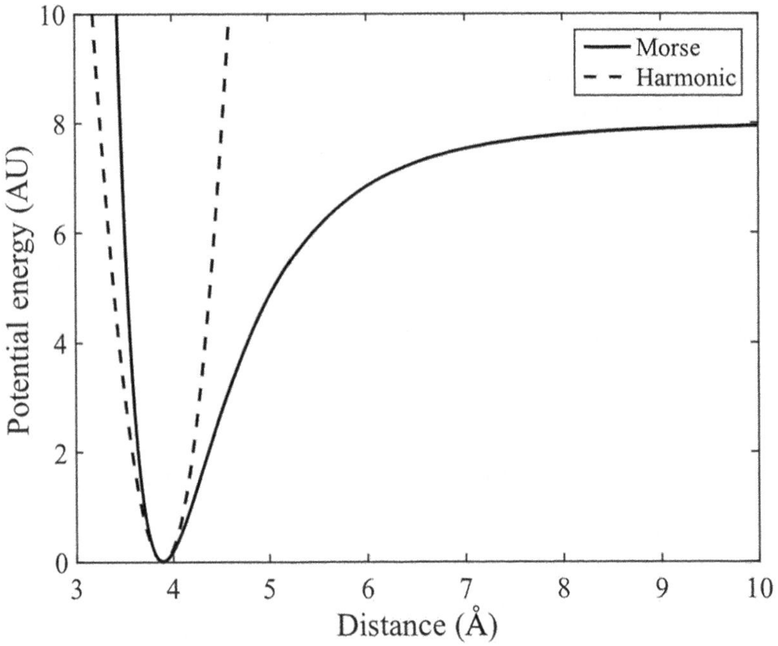

Figure 4.9 Morse and harmonic potentials.

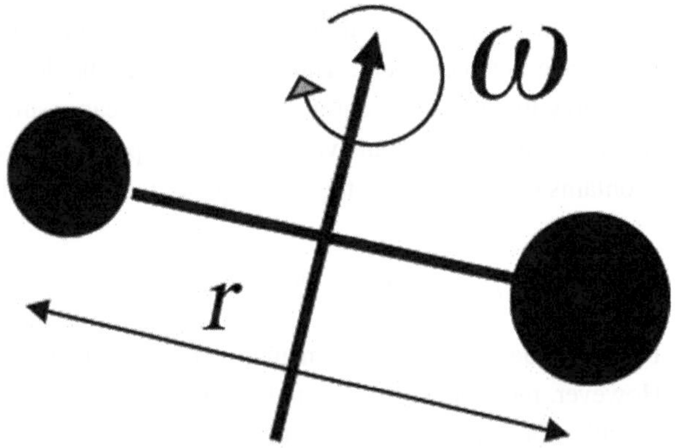

Figure 4.10 Rotational motion of a diatomic molecule.

nuclei are rigidly affixed to each other, then we can use the angular solution to the wave equation that we developed above. This is called the rigid rotor approximation. The angular momentum is

$$L = I\omega \tag{4.89}$$

where I is the moment of inertia

$$I = \mu r^2 \tag{4.90}$$

and μ is the reduced mass

$$\mu = \frac{m_1 \times m_2}{m_1 + m_2} \tag{4.91}$$

The energy of rotation then becomes

$$\epsilon_R = \frac{1}{2} I\omega^2 = \frac{L^2}{2I} = l(l+1)\frac{\hbar^2}{2\mu r^2} \tag{4.92}$$

For molecular rotation it is common to use J as the rotational quantum number, so

$$\epsilon_R = J(J+1)\frac{\hbar^2}{2\mu r^2} \tag{4.93}$$

Next we consider vibration, or the relative motion of nuclei with respect to each other (Fig. 4.11). If we neglect rotation and consider that the intermolecular force acts like a linear spring, then we can use the harmonic potential of Eq. (4.9) in the radial wave equation

$$\frac{\hbar^2}{2\mu r^2}\frac{d}{dr}\left(r^2\frac{dR}{dr}\right) + \left\{\epsilon - V(r) - \frac{\hbar^2}{2\mu r^2} l(l+1)\right\} R = 0 \tag{4.94}$$

This equation can be solved analytically using series methods. Introducing the transformations

$$K(r) = rR(r) \text{ and } x = r - r_e \tag{4.95}$$

the equation becomes

$$\frac{dK^2}{dx^2} + \frac{2\mu}{\hbar^2}\left(\epsilon_V - \frac{kx^2}{2}\right)K = 0$$

Introducing a non-dimensional form for energy (where ν is a characteristic frequency)

$$\lambda = \frac{2\epsilon_V}{h\nu_V}, \text{ where } \nu_V = \frac{1}{2\pi}\sqrt{\frac{k}{\mu}} \tag{4.96}$$

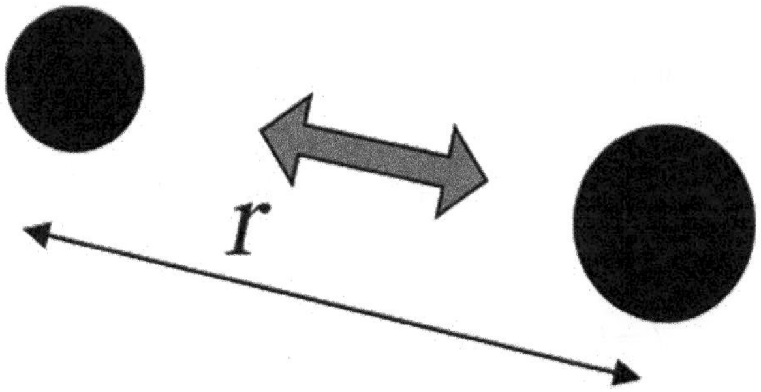

Figure 4.11 Vibrational motion of a diatomic molecule.

and defining a new independent variable

$$y = \left(\frac{2\pi \nu_V \mu}{\hbar}\right)^{1/2} x \tag{4.97}$$

one obtains a non-dimensional form of the radial wave equation

$$\frac{d^2K}{dy^2} + (\lambda - y^2)K = 0 \tag{4.98}$$

Let us examine the solution as $y \to \infty$:

$$\frac{d^2K}{dy^2} - y^2K \cong 0 \Rightarrow K \cong \exp(-y^2/2) \tag{4.99}$$

Assume that we can express $K(y)$ as a factor $H(y)$ times the large y limiting solution

$$K(y) = H(y)\exp(-y^2/2) \tag{4.100}$$

Then

$$\frac{d^2H}{dy^2} - 2y\frac{dH}{dy} + (\lambda - 1)H = 0 \tag{4.101}$$

This is another equation solved by a nineteenth-century mathematician, Charles Hermite. It has solutions only when

$$(\lambda - 1) = 2v,\ v = 0, 1, 2, 3, \ldots \tag{4.102}$$

$H(y)$ is the Hermite polynomial of degree u:

$$H_u(y) = (-1)^u e^{y^2} \frac{d^u}{dy^u}\left(e^{-y^2}\right) \tag{4.103}$$

The energy is

$$\epsilon_v = \left(v + \frac{1}{2}\right) h\nu_v, \text{ where } v = 0, 1, 2, 3, \ldots \tag{4.104}$$

Note that energy does NOT go to zero. $\epsilon_v = \frac{1}{2}h\nu_v$ is called the zero-point energy. In addition, the degeneracy is unity ($g_v = 1$). The resulting vibrational energy levels are shown superimposed on Fig. 4.12, along with what the real energy levels might look like.

Putting the rotational and vibrational solutions together, the total energy of these two modes becomes

$$\epsilon_R + \epsilon_v = \frac{\hbar^2}{2I_e}J(J+1) + h\nu_v\left(v + \frac{1}{2}\right) \tag{4.105}$$

where

$$J = 0, 1, 2, 3 \ldots \text{ and } v = 0, 1, 2, 3 \ldots \tag{4.106}$$

with degeneracies

$$g_J = 2J + 1 \text{ and } g_v = 1 \tag{4.107}$$

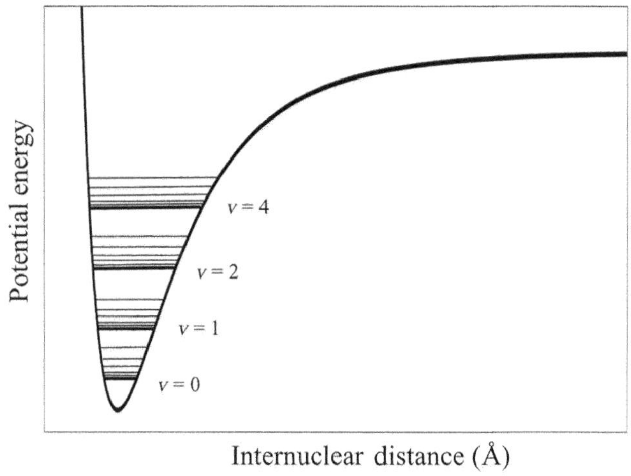

Figure 4.12 Rotational and vibrational energy levels for a diatomic molecule.

The energy is more commonly expressed as

$$F(J) = \frac{\epsilon_R}{hc} = B_e J(J+1) \text{ and } G(\mathrm{v}) = \frac{\epsilon_v}{hc} = \omega_e \left(\mathrm{v} + \frac{1}{2}\right) \tag{4.108}$$

where the units are cm^{-1}. This nomenclature was developed by spectroscopists, who noted that the inverse of the wavelength of spectroscopic transitions is proportional to the energy difference between the two quantum states involved in the transition because

$$\epsilon = h\nu = h\frac{c}{\lambda} \tag{4.109}$$

Alternatively, we can express the energy in terms of characteristic temperatures:

$$T_R \equiv \frac{\hbar^2}{2I_e k} = \frac{B_e}{(k/hc)} \text{ and } T_V \equiv \frac{h\nu_v}{k} = \frac{G_v}{(k/hc)} \tag{4.110}$$

For N_2, for example,

$$B_e = 1.998\ \mathrm{cm}^{-1}, \qquad T_R = 2.87\ \mathrm{K}$$

$$\omega_e = 2357.6\ \mathrm{cm}^{-1}, \qquad T_V = 3390\ \mathrm{K}$$

Note that $B_e << \omega_e$. Thus, the rotational energy levels are much more closely spaced than the vibrational levels. This is illustrated in Fig. 4.12, although the rotational levels for a real molecule are much more closely spaced than shown in this cartoon. (A useful conversion relation is 1 ev = 1.602×10^{-19} J = 11,600 K = 8058 cm^{-1} = 23.06 kcal/mol.)

Next we explore real atomic and molecular behavior.

4.4 Real Atomic Behavior

Unfortunately, real atoms (including the hydrogen atom) are not very well described by the simple theory. The following contribute to actual structure:

- Quantum mechanics of identical particles: the Pauli exclusion principle.
- Higher-order effects:
  - spin–orbit coupling
  - relativistic
  - nuclear spin
  - Zeman effect.
- Multiple electrons.

4.4.1 Pauli Exclusion Principle

The Pauli exclusion principle states that no two electrons of a multi-electron atom can assume the same quantum state. This comes about because of symmetry considerations when dealing with multiple particles. Consider, for example, the wave equation for two identical particles in a box:

$$-\frac{\hbar^2}{2m}\left(\frac{\partial\psi^2}{\partial^2 x_1}+\frac{\partial\psi^2}{\partial^2 x_2}\right)+V\psi=0 \tag{4.111}$$

where

$$\begin{aligned}\psi &= \psi(x_1,x_2)\\ V &= V(x_1,x_2)\end{aligned} \tag{4.112}$$

If the particles are interchanged, the probability density of the system should be unchanged. This means

$$p(1,2)=p(2,1) \tag{4.113}$$

Since

$$p(\vec{r},t)dV=\Psi^*(\vec{r},t)\Psi(\vec{r},t)dV=|\Psi(\vec{r},t)|^2 \tag{4.114}$$

for the probability to be unchanged upon exchange we must have

$$\psi(1,2)=\psi(2,1) \tag{4.115}$$

for a symmetric wave function and

$$\psi(1,2)=-\psi(2,1) \tag{4.116}$$

for an anti-symmetric wave function.

Particles with half-integer spin quantum number have anti-symmetric wave functions and are called fermions, after the famous physicist Enrico Fermi. Electrons, protons, neutrons, and atoms and molecules with an odd number of electrons, protons, and neutrons are fermions. Fermions obey Fermi–Dirac statistics. Particles with integer spin quantum number have symmetric wave functions and are called bosons, after the Indian physicist Satyendra Bose. Photons and atoms and molecules with an even number of electrons, protons, and neutrons are bosons. Bosons obey Bose–Einstein statistics.

Assume we can write the potential energy function as

$$V = V(x_1) + V(x_2) = V(1) + V(2) \tag{4.117}$$

then

$$\psi_{\alpha,\beta}(1,2) = \psi_\alpha(1)\psi_\beta(2) \tag{4.118}$$

where α and β denote the quantum state of each particle. We can now construct a wave function that satisfies the symmetry requirement:

$$\psi_{s,a}(1,2) = \frac{1}{\sqrt{2}}\left\{\psi_\alpha(1)\psi_\beta(2) \pm \psi_\alpha(2)\psi_\beta(1)\right\} \tag{4.119}$$

This clearly satisfies $\psi(1,2) = \pm\psi(2,1)$. Note, however, that if the two particles are in the same quantum state, and the wave function is anti-symmetric, then $\psi \rightarrow 0$. Thus, fermions cannot occupy the same quantum state at the same time. Hence, the Pauli exclusion principle and the structure of all multi-electron atoms derive from this fact.

4.4.2 Higher-Order Effects

4.4.2.1 Spin–Orbit Coupling

Electrons have spin, and thus angular momentum, as does the entire atom. These two motions generate magnetic fields that interact, thus changing the energies of individual quantum states. The total angular momentum vector is

$$\vec{J} = \vec{L} + \vec{S} \tag{4.120}$$

One can show that

$$J = \sqrt{j(j+1)}\hbar \text{ and } J_z = m_j\hbar \tag{4.121}$$

where

$$\begin{aligned} &m_j = -j, -j+1, \ldots, j-1, j \\ &l = 0:\ j = 1/2 \\ &l \neq 0:\ j = l \pm 1/2 \end{aligned} \tag{4.122}$$

Thus, m_l and m_s cease to properly describe the quantum structure and are replaced by j and m_j. (m_l and m_s are said to be bad quantum numbers.) Good quantum numbers are therefore

$$n,\ l,\ j,\ m_j \tag{4.123}$$

Spin–orbit coupling causes quantum states that would otherwise have the same energy to split, thus changing the degeneracy as well, sometimes by large amounts. This is typically characterized using the following relation:

$$\epsilon = \epsilon_n + \Delta\epsilon_{SL} \tag{4.124}$$

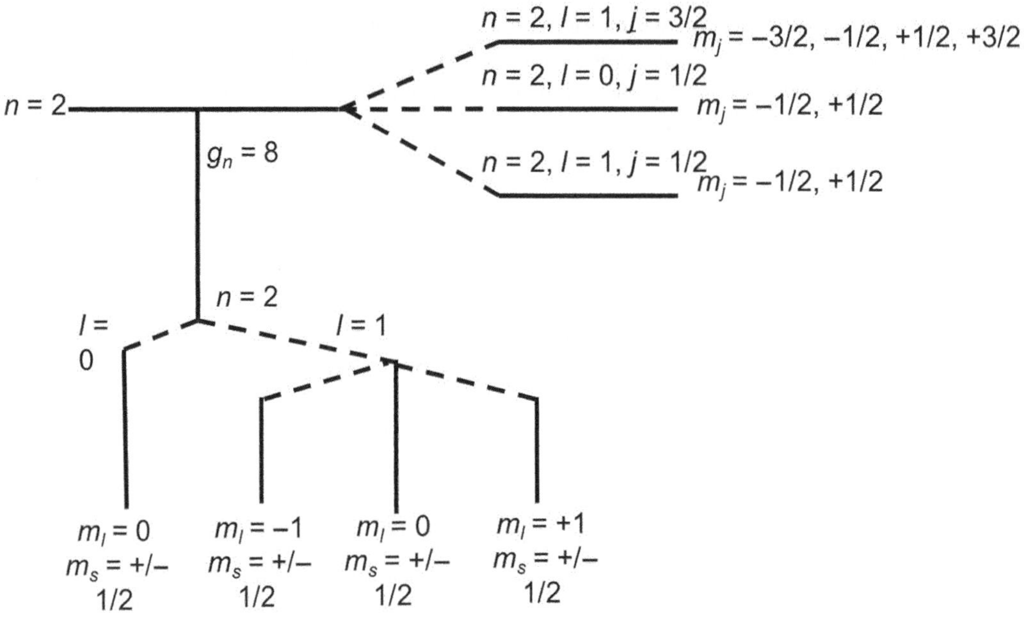

Figure 4.13 Fine structure splitting for a hydrogenic atom (Incropera [7]).

where

$$l = 0: \ \Delta\epsilon_{SL} = 0$$

$$l \neq 0 \begin{cases} j = l + 1/2: \ \Delta\epsilon_{SL} = \frac{Z^2\alpha^2}{n(2l+1)(l+1)} |\epsilon_n| \\ j = l - 1/2: \ \Delta\epsilon_{SL} = \frac{Z^2\alpha^2}{nl(2l+1)} |\epsilon_n| \end{cases} \tag{4.125}$$

Here α is called the fine structure constant and is equal to

$$\alpha = \frac{e^2}{\hbar c} = \frac{1}{137} \tag{4.126}$$

Figure 4.13 gives an example for a hydrogenic atom. (Note that the total degeneracy does not change.)

4.4.2.2 Relativistic Effects

If we included relativistic effects we would find that they change the quantum state energies by an amount similar to spin–orbit splitting and approximately equal to

$$\Delta\epsilon_R = -\frac{Z^2\alpha^2}{en^2}\left(\frac{4n}{l + 1/2} - 3\right)|\epsilon_n| \tag{4.127}$$

4.4.2.3 Nuclear Spin

As with electrons, nuclei have angular momentum or spin. Spin induces a magnetic dipole moment that interacts with orbital angular momentum, although to a much lesser degree than electron spin. This interaction leads to a splitting of otherwise degenerate quantum levels. The resulting splits are called the hyperfine structure, and from a thermodynamics point of view are usually ignored.

4.4.2.4 The Zeeman Effect

If an atom is placed in an external magnetic field **H**, the quantum states will be altered in a way that can cause level splitting. This is known as the Zeeman effect. The size of the splitting depends on the orientation of the total orbital angular momentum vector **J** and **H**, and therefore depends on the magnetic quantum number m_j. The energy splitting is given by

$$\Delta\epsilon_z = \frac{e\hbar}{2m_e c} g m_j H \tag{4.128}$$

where g is the Lande g factor (approximately 2.0023).

4.4.3 Multiple Electrons

For multi-electron atoms, the solution of the wave equation is more complex. However, optical spectra in atoms originate from the interaction of radiation with the outermost electrons of the atom, called the optical or valence electrons. The behavior of these electrons with respect to the more tightly bound inner electrons and nucleus is somewhat similar to that of a single electron and a nucleus. It is usual to characterize electrons by the type of orbit they occupy, according to the energy and the orbital angular momentum. The first is specified by the principal quantum number, and the second by a letter which corresponds to the angular momentum. The letters are:

$$\begin{array}{cccccc} 0 & 1 & 2 & 3 & 4 & \ldots \\ s & p & d & f & g & \ldots \end{array}$$

(My mnemonic for remembering this is "San Police Department Frisco!")

The orbits of the electrons are referred to as $3s^2$, $2p^4$, and so on, where the superscript refers to the number of electrons in the orbits. Thus, the structure of a given atom can be given as a list of the orbits of its electrons. Pauli's exclusion principle limits the number of electrons that occupy a given orbit at the same time to $2(2l{+}1)$. The minimum-energy configuration of an atom is called its ground state. The ground state can be characterized by filling in the allowed orbits in energy order with no more than $2(2l+1)$ electrons per orbit. As an example, the oxygen atom ground-state electron configuration is

$$1s^2 2s^2 2p^4 \tag{4.129}$$

Note that the sum of the superscripts must equal the number of electrons, eight in the case of oxygen. A detailed table of atomic electron configurations and other useful information is given in Table C.1 (Appendix C). Note that as electrons are added, there will be elements whose outermost orbit has exactly $2(2l+1)$ electrons. These elements, called the noble gases, are unusually stable. This manifests itself in high ionization potentials, as seen in the table. On the other hand, those elements with one electron in the outermost orbit are much less stable. They are called the alkali metals. These structural differences are illustrated in the periodic table (Fig. 4.14).

Periodic Table of the Elements
Ground State Electron Configurations

http://chemistry.about.com
©2012 Todd Helmenstine
About Chemistry

1A	2A	3B	4B	5B	6B	7B	8B	8B	8B	1B	2B	3A	4A	5A	6A	7A	8A
1 **H** $1s^1$																	2 **He** $1s^2$
3 **Li** $1s^2 2s^1$	4 **Be** $1s^2 2s^2$											5 **B** $1s^2 2s^2 p^1$	6 **C** $1s^2 2s^2 p^2$	7 **N** $1s^2 2s^2 p^3$	8 **O** $1s^2 2s^2 p^4$	9 **F** $1s^2 2s^2 p^5$	10 **Ne** $1s^2 2s^2 p^6$
11 **Na** [Ne]$3s^1$	12 **Mg** [Ne]$3s^2$											13 **Al** [Ne]$3s^2 p^1$	14 **Si** [Ne]$3s^2 p^2$	15 **P** [Ne]$3s^2 p^3$	16 **S** [Ne]$3s^2 p^4$	17 **Cl** [Ne]$3s^2 p^5$	18 **Ar** [Ne]$3s^2 p^6$
19 **K** [Ar]$4s^1$	20 **Ca** [Ar]$4s^2$	21 **Sc** [Ar]$3d^1 4s^2$	22 **Ti** [Ar]$3d^2 4s^2$	23 **V** [Ar]$3d^3 4s^2$	24 **Cr** [Ar]$3d^5 4s^1$	25 **Mn** [Ar]$3d^5 4s^2$	26 **Fe** [Ar]$3d^6 4s^2$	27 **Co** [Ar]$3d^7 4s^2$	28 **Ni** [Ar]$3d^8 4s^2$	29 **Cu** [Ar]$3d^{10} 4s^1$	30 **Zn** [Ar]$3d^{10} 4s^2$	31 **Ga** [Ar]$3d^{10} 4s^2 p^1$	32 **Ge** [Ar]$3d^{10} 4s^2 p^2$	33 **As** [Ar]$3d^{10} 4s^2 p^3$	34 **Se** [Ar]$3d^{10} 4s^2 p^4$	35 **Br** [Ar]$3d^{10} 4s^2 p^5$	36 **Kr** [Ar]$3d^{10} 4s^2 p^6$
37 **Rb** [Kr]$5s^1$	38 **Sr** [Kr]$5s^2$	39 **Y** [Kr]$4d^1 5s^2$	40 **Zr** [Kr]$4d^2 5s^2$	41 **Nb** [Kr]$4d^4 5s^1$	42 **Mo** [Kr]$4d^5 5s^1$	43 **Tc** [Kr]$4d^5 5s^2$	44 **Ru** [Kr]$4d^7 5s^1$	45 **Rh** [Kr]$4d^8 5s^1$	46 **Pd** [Kr]$4d^{10}$	47 **Ag** [Kr]$4d^{10} 5s^1$	48 **Cd** [Kr]$4d^{10} 5s^2$	49 **In** [Kr]$4d^{10} 5s^2 p^1$	50 **Sn** [Kr]$4d^{10} 5s^2 p^2$	51 **Sb** [Kr]$4d^{10} 5s^2 p^3$	52 **Te** [Kr]$4d^{10} 5s^2 p^4$	53 **I** [Kr]$4d^{10} 5s^2 p^5$	54 **Xe** [Kr]$4d^{10} 5s^2 p^6$
55 **Cs** [Xe]$6s^1$	56 **Ba** [Xe]$6s^2$	57-71 Lanthanides	72 **Hf** [Xe]$4f^{14} 5d^2 6s^2$	73 **Ta** [Xe]$4f^{14} 5d^3 6s^2$	74 **W** [Xe]$4f^{14} 5d^4 6s^2$	75 **Re** [Xe]$4f^{14} 5d^5 6s^2$	76 **Os** [Xe]$4f^{14} 5d^6 6s^2$	77 **Ir** [Xe]$4f^{14} 5d^7 6s^2$	78 **Pt** [Xe]$4f^{14} 5d^9 6s^1$	79 **Au** [Xe]$4f^{14} 5d^{10} 6s^1$	80 **Hg** [Xe]$4f^{14} 5d^{10} 6s^2$	81 **Tl** [Xe]$4f^{14} 5d^{10} 6s^2 p^1$	82 **Pb** [Xe]$4f^{14} 5d^{10} 6s^2 p^2$	83 **Bi** [Xe]$4f^{14} 5d^{10} 6s^2 p^3$	84 **Po** [Xe]$4f^{14} 5d^{10} 6s^2 p^4$	85 **At** [Xe]$4f^{14} 5d^{10} 6s^2 p^5$	86 **Rn** [Xe]$4f^{14} 5d^{10} 6s^2 p^6$
87 **Fr** [Rn]$7s^1$	88 **Ra** [Rn]$7s^2$	89-103 Actinides	104 **Rf** [Rn]$5f^{14} 6d^2 7s^2$	105 **Db** [Rn]$5f^{14} 6d^3 7s^2$	106 **Sg** [Rn]$5f^{14} 6d^4 7s^2$	107 **Bh** [Rn]$5f^{14} 6d^5 7s^2$	108 **Hs** [Rn]$5f^{14} 6d^6 7s^2$	109 **Mt** [Rn]$5f^{14} 6d^7 7s^2$	110 **Ds** [Rn]$5f^{14} 6d^8 7s^2$	111 **Rg** [Rn]$5f^{14} 6d^9 7s^2$	112 **Cn** [Rn]$5f^{14} 6d^{10} 7s^2$	113 **Uut**	114 **Fl**	115 **Uup**	116 **Lv**	117 **Uus**	118 **Uuo**

	57 **La** [Xe]$5d^1 6s^2$	58 **Ce** [Xe]$4f^1 5d^1 6s^2$	59 **Pr** [Xe]$4f^3 6s^2$	60 **Nd** [Xe]$4f^4 6s^2$	61 **Pm** [Xe]$4f^5 6s^2$	62 **Sm** [Xe]$4f^6 6s^2$	63 **Eu** [Xe]$4f^7 6s^2$	64 **Gd** [Xe]$4f^7 5d^1 6s^2$	65 **Tb** [Xe]$4f^9 6s^2$	66 **Dy** [Xe]$4f^{10} 6s^2$	67 **Ho** [Xe]$4f^{11} 6s^2$	68 **Er** [Xe]$4f^{12} 6s^2$	69 **Tm** [Xe]$4f^{13} 6s^2$	70 **Yb** [Xe]$4f^{14} 6s^2$	71 **Lu** [Xe]$4f^{14} 5d^1 6s^2$
Lanthanides															
Actinides	89 **Ac** [Rn]$6d^1 7s^2$	90 **Th** [Rn]$6d^2 7s^2$	91 **Pa** [Rn]$5f^2 6d^1 7s^2$	92 **U** [Rn]$5f^3 6d^1 7s^2$	93 **Np** [Rn]$5f^4 6d^1 7s^2$	94 **Pu** [Rn]$5f^6 7s^2$	95 **Am** [Rn]$5f^7 7s^2$	96 **Cm** [Rn]$5f^7 6d^1 7s^2$	97 **Bk** [Rn]$5f^9 7s^2$	98 **Cf** [Rn]$5f^{10} 7s^2$	99 **Es** [Rn]$5f^{11} 7s^2$	100 **Fm** [Rn]$5f^{12} 7s^2$	101 **Md** [Rn]$5f^{13} 7s^2$	102 **No** [Rn]$5f^{14} 7s^2$	103 **Lr** [Rn]$5f^{14} 6d^1 7s^2$

* values are based on theory and are not verified

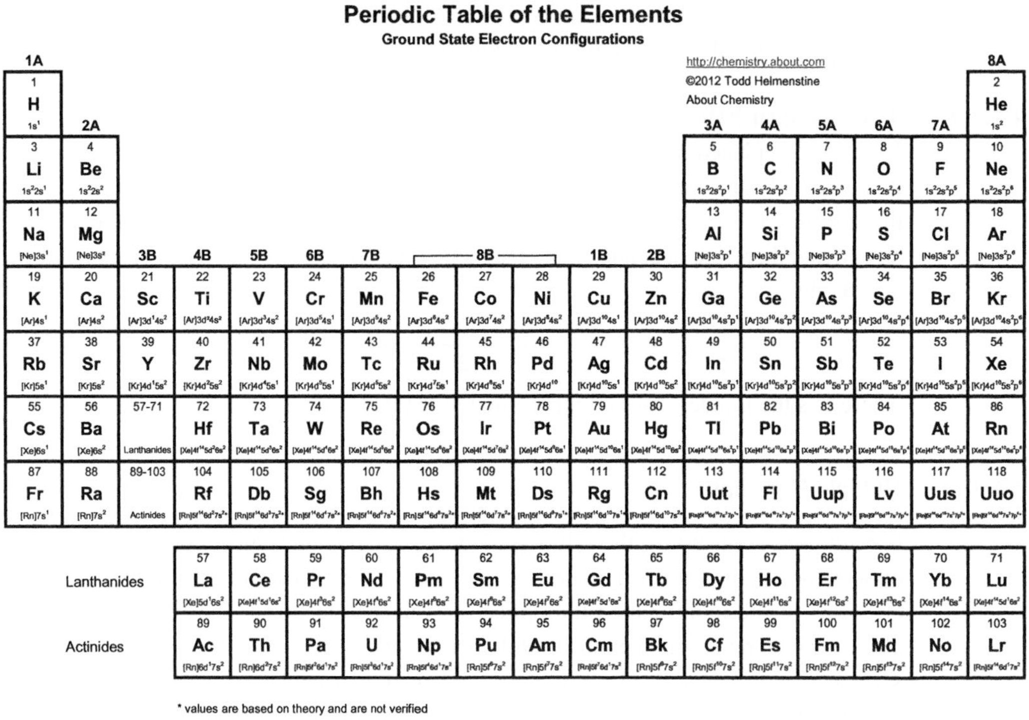

Figure 4.14 The periodic table.

The possible angular momentum states of the atom are described by the spectroscopic term classification. The energy levels of most atoms can be described in terms of L, the total orbital angular momentum quantum number, S, the total spin angular momentum, and J, the total angular momentum quantum number. There may be several different values of J for a given L and S. These are

$$J = (L + S), (L + S - 1), (L + S - 2), \ldots, |L - S| \qquad (4.130)$$

Thus, $2S + 1$ values of J will result if $S < L$ and $2L + 1$ values if $S > L$. The term symbol, or classification, is of the form

$$^{2S+1}L_J \qquad (4.131)$$

L is designated by the following capital letter notation:

$$
\begin{array}{ccccccc}
0 & 1 & 2 & 3 & 4 & \ldots \\
S & P & D & F & G & \ldots
\end{array}
$$

An important property is the multiplicity. This is defined as the number of different values of J which are possible for given values of L and S. It is equal to $2S + 1$ for $S \leq L$ and $2L + 1$ for $L < S$. The superscript on the term classification is equal to the multiplicity only if $S \leq L$. A multiplet is a group of quantum states that are in every way

Table 4.2 Electronic energy levels of sodium

Optical electron configuration	Term classification	Energy (cm^{-1})	Degeneracy
$3s$	$^2S_{1/2}$	0	2
$3p$	$^2P_{1/2,3/2}$	16,965	6
$4s$	$^2S_{1/2}$	25,740	2
$3d$	$^2D_{5/2,3/2}$	29,173	10
$4p$	$^{2a}P_{1/2,3/2}$	30,269	6
$5s$	$^2S_{1/2}$	33,201	2

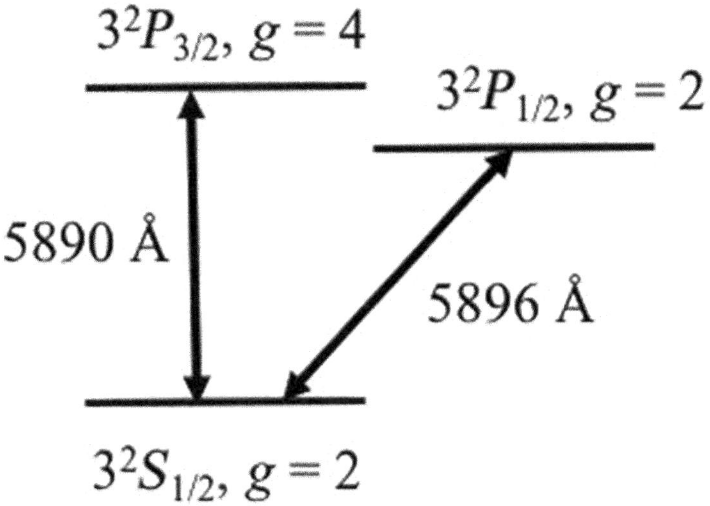

Figure 4.15 Sodium energy-level diagram.

equal, except for differing energy levels. Because of spin–orbit coupling, these states may have slightly different energies. The degeneracy of an energy level with given J is

$$g_J = 2J + 1 \tag{4.132}$$

If the energies of all components of a multiplet are equal, the degeneracy of the multiplet is

$$g_{mult} = \sum_{i=1}^{N} g_{J,i} \tag{4.133}$$

where N is the number of components in the multiplet.

Data on atomic structure are typically presented in tabular and graphical form. An example of the lower electronic energy levels for sodium is given in Table 4.2. The energy levels can be represented graphically using an energy-level diagram. An example for sodium is given in Fig. 4.15. Note the $^2P_{1/2,3/2}$ multiplet. The radiative transitions to these two components from the $^2S_{1/2}$ state are the famous Fraunhofer D lines.

4.5 Real Molecular Behavior

All of the above complexities, such as spin–orbit coupling, carry through to molecular behavior.

The electronic quantum state is characterized by the axial orbital angular momentum quantum number Λ, the total electron spin quantum number S, and the axial spin quantum number Σ. Λ can take on values of $0, 1, 2, \ldots$ S may assume integer or half-integer values, and Σ can take on values

$$\Sigma = -S, -S+1, \ldots, S-1, S \tag{4.134}$$

The term classification symbol for the diatomic molecule is

$$^{2S+1}\Lambda_{\Lambda+\Sigma} \tag{4.135}$$

As with atoms, the value of Λ is represented by letters, in this case Greek:

0	1	2	3	4	...
Σ	Π	Δ	Φ	Γ	...

Multiplicity is also an important property in molecules. For states with $\Lambda > 0$, orbital–spin interactions can cause small variations in energy for states with different Σ. The number of such states is equal to $2S$+1. Thus, the degeneracy of each level is

$$g_e = 2S + 1 \tag{4.136}$$

for $\Lambda = 0$ and

$$g_e = 2 \tag{4.137}$$

for $\Lambda > 0$. As an example, the state classifications for the lowest four electronic states of NO are

Classification	Energy	g_e
$^2\Pi_{1/2}$	0	2
$^2\Pi_{3/2}$	121	2
$^2\Sigma^+$	43,965	2
$^2\Pi$	45,930	4

In most molecules, the potential energy function is not exactly harmonic nor is the rigid rotator approximation exactly met. For small departures from ideal conditions, a perturbation analysis is appropriate. In this case, the energy terms become

$$\begin{aligned} G(v) + F_v(J) \cong\ & \omega_e(v+1/2) - \omega_e x_e(v+1/2)^2 + \omega_e y_e(v+1/2)^3 \\ & - \cdots + B_v J(J+1) - D_v J^2(J+1)^2 + \cdots \end{aligned} \tag{4.138}$$

where x_e, y_e, and D_v are additional constants depending on the particular molecule and its electronic state. Data for a wide variety of diatomic molecules were tabulated by Herzberg [28] and can be found in the NIST Chemistry Webbook.

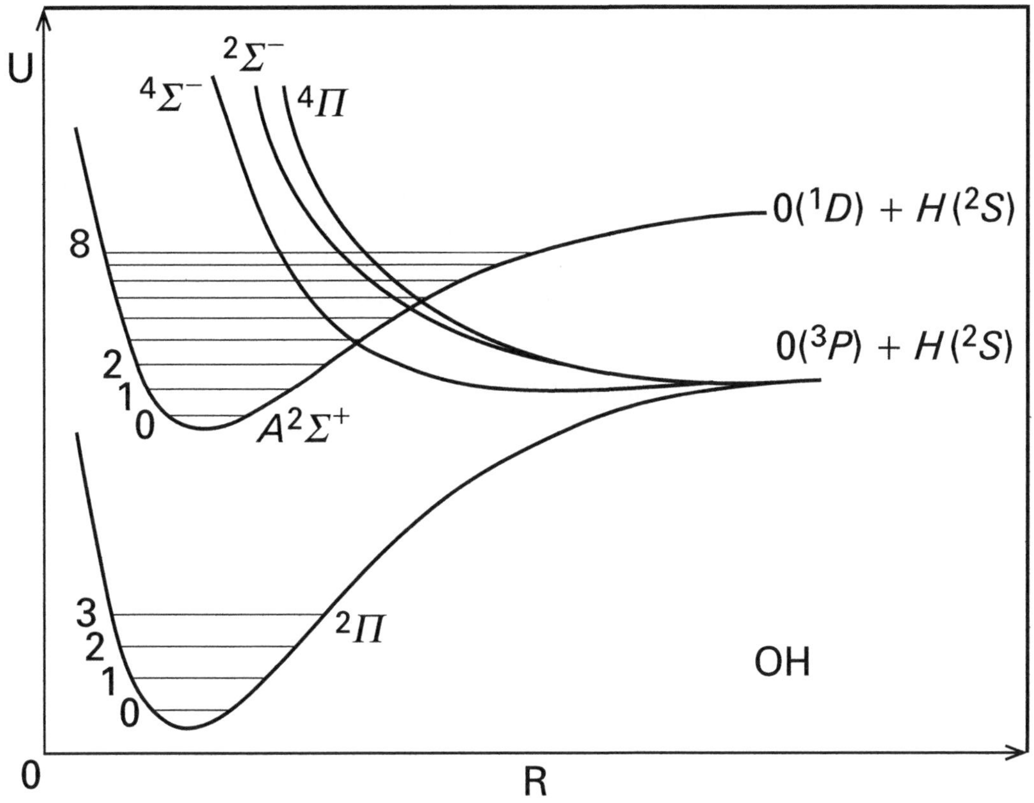

Figure 4.16 OH energy-level diagram (from Radzig and Smirov [9]).

As with atoms, molecular energy states can be represented graphically. However, it is usual to plot the energy levels in terms of the electronic potential functions as a function of separation of the nuclei. The vibrational and rotational states are then plotted as horizontal lines on each potential. As an example, Fig. 4.16 shows the energy level diagram for OH. Not all electronic configurations result in stable potentials. The $^4\Sigma^-$ state is an example of an unstable state, in that it always results in a repulsive force between the two nuclei. Such states are called predissociative. There are situations in which such states can be used in laser-induced fluorescence instrumentation schemes.

As the number of nuclei increases, the complexity of describing their structure increases correspondingly. In a molecule with n atoms, there are $3n$ degrees of freedom of motion. Three degrees are associated with translational motion. For the general case, three degrees are associated with the angular orientation of the molecule, or rotational motion. Thus, in general, there are $3n - 6$ degrees of freedom for vibrational motion. The exception are linear molecules which have only two degrees of rotational freedom, and thus $3n - 5$ degrees of vibrational freedom. In some cases, the Born–Oppenheimer approximation holds, and simple electronic potentials result. In this case, the vibrational and rotational expressions given above are descriptive. In many cases, however, the complexity prevents this simple view. See Herzberg [29] for a detailed discussion of the structure and spectral properties of polyatomic molecules.

In practice, formulas such as those given by Eq. (4.138) are sometimes not sufficiently accurate for precision spectroscopic purposes. The approximations made in their derivation do not hold exactly, and the resulting errors can result in incorrect identification of spectral features. As a result, more precise structural data are usually required. One compilation of such data is LIFBASE, a program written by Luque and Crosley [30] at the Stanford Research Institute to assist in interpreting laser-induced fluorescence measurements. The program contains data for OH, OD, NO, CH, and CN.

4.6 Molecular Modeling/Computational Chemistry

Molecular modeling or computational chemistry is the discipline of using numerical methods to model atomic and molecular behavior. It is used to study structure, energies, electron density distribution, moments of the electron density distribution, rotational and vibrational frequencies, reactivity, and collision dynamics. It can be used to model individual atoms or molecules, gases, liquids, and solids. Methods range from the highly empirical to *ab initio*, which means from basic principles. *Ab initio* methods are based on quantum mechanics and fundamental constants. Other methods are called semi-empirical as they depend on parametrization using experimental data.

While the simple solutions to the wave equation that we derived in Section 4.3 are extremely useful for understanding the underlying physics of atomic and molecular structure, they are not very accurate. Two good examples are the rigid rotator and the harmonic oscillator. A key assumption of the rigid rotator model is that the distance between the atoms is fixed. In fact, it depends on the vibrational state, thus distorting the simple solution. The harmonic oscillator ignores non-harmonic behavior as the molecule approaches the dissociation state. Even the polynomial correction formulas given in the previous section are not particularly accurate if fine details are desired, or as the potential function becomes more asymmetric.

Molecular modeling methods include the main branches listed in Table 4.3.

The reason that different methods have evolved has to do with computational efficiency and the need for certain levels of accuracy. The least-expensive methods are

Table 4.3 Methods of molecular modeling

Method	Description
Molecular mechanics/molecular dynamics	Modeling the motion of atoms and molecules using Newton's Law. Forces between atoms are modeled algebraically.
Hartree–Fock (HF)	Numerical solution of the wave equation. There are many variations of HF.
Density functional theory (DFT)	Based on the concept of density functionals. Less accurate than HF methods but computationally less expensive.
Semi-empirical	Methods that parameterize some portion of a HF or DFT model to speed calculation.

Table 4.4 Molecular modeling software

Gaussian	General-purpose commercial software package. Includes MM, HF and enhancements, DFT, QM/MM, and other methods.
GAMESS	Open source *ab initio* quantum chemistry package.
Amsterdam	Commercial DFT and reactive MD package.
PSI	Open source HF and enhancements package.

those based on molecular mechanics, which we discuss in more detail in Chapter 9 on liquids. Necessarily, they are the least accurate. At the other limit, Hartree–Fock methods and their more advanced forms, which include electron correlations or coupled-cluster theory, offer the highest accuracy, but at very high computational cost. Indeed, typically they are limited to systems with no more than 10 or so atoms. Semi-empirical methods were developed to be less costly than full *ab initio* methods, while maintaining some of the benefits of the full methods. DFT methods, while less accurate than HF methods, are much more accurate that semi-empirical methods and mostly used to study solid-state behavior. There are also combined methods, such as QM/MM (quantum mechanics/molecular mechanics), which are used to model reactions between large biomolecules where the reaction involves only a few atoms.

There are many commercial and open source computational chemistry software packages available. Table 4.4 provides a short list of packages that find common use in engineering. Gaussian is probably the most popular commercial package for those needing HF and post-HF methods, although it covers all four major method areas. Amsterdam is mostly used by materials researchers. GAMESS is one of the most used open source *ab initio* codes. PSI is a simpler open source alternative to GAMESS.

An important consideration is how the user is to interact with the calculations. The output of these codes involve large sets of data, which are difficult to parse. Therefore, it is necessary to have some form of visualization capability. Gaussian and Amsterdam come with their own visualization software, while GAMESS and PSI require third-party packages, of which fortunately there are many. VMD and Avogadro are two examples.

4.6.1 Example 4.1

Calculate the equilibrium structure of the molecule CH_2 using GAMESS and plot the result using VMD. Methylene is a radical species, meaning it would rather be CH_4 and so is highly reactive.

Below is an input file for this calculation.

Listing 4.1: GAMESS input for CH_2

```
 $CONTRL SCFTYP=RHF RUNTYP=OPTIMIZE COORD=ZMT NZVAR=0 $END
 $SYSTEM TIMLIM=1 $END
 $STATPT OPTTOL=1.0E-5  $END
 $BASIS  GBASIS=STO NGAUSS=2 $END
 $GUESS  GUESS=HUCKEL $END
 $DATA
```

```
Methylene...1 -A-1 state ...RHF/STO-2G
Cnv  2

C
H  1 rCH
H  1 rCH  2 aHCH

rCH=1.09
aHCH=110.0
 $END
```

The output file is quite large. However, it can be read by VMD or Avogadro. Doing so, the predicted structure can be displayed and the geometry measured. In this case the bond distances are each 1.124 Å and the bond angle is 99.3°. See Fig. 4.17.

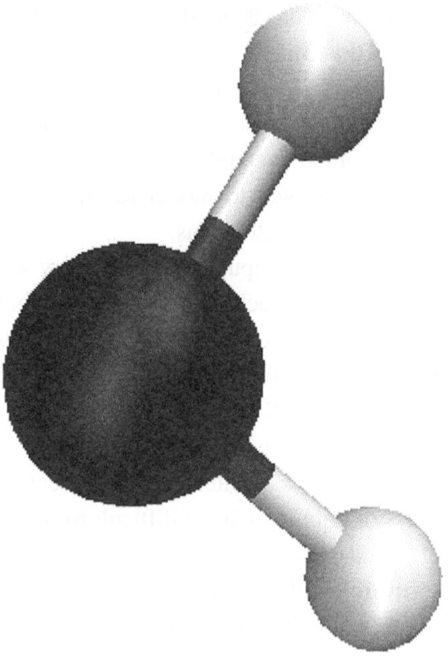

Figure 4.17 Predicted structure of CH_2.

4.7 Summary

The most important result of quantum mechanics is that atoms and molecules typically exist in discrete states, known as quantum states. It turns out that at the very small length and time scales of atomic and molecular behavior, Newtonian physics do not apply. Instead, their behavior is the domain of quantum mechanics. In this chapter we provided a brief history of how quantum mechanics came to be developed, including the wave–particle duality question and Heisenberg's uncertainty principle. We then introduced the Schrödinger wave equation:

Table 4.5 Simple solutions of the wave equation

	Quantum number	Energy	Degeneracy
Particle in box	n_x, n_y, n_z	$\epsilon_{tr} = \frac{h^2}{8mV^{2/3}}(n_x^2 + n_y^2 + n_z^2)$	(see Chapter 5)
Rigid rotator	J	$\epsilon_R = J(J+1)B_e$	$g_J = 2J + 1$
Harmonic oscillator	v	$\epsilon_V = \omega_e(v + 1/2)$	$g_v = 1$
Hydrogenic atom	n	$\epsilon_n = -\frac{Z^2 e^4 \mu}{32\pi^2\epsilon_0^2\hbar^2}\frac{1}{n^2}$	$g_n = 2n^2$

$$-\frac{\hbar^2}{2m}\nabla^2\Psi(\vec{r},t) + V(\vec{r},t)\Psi(\vec{r},t) = i\hbar\frac{\partial\Psi(\vec{r},t)}{\partial t} \tag{4.139}$$

We explored some simple solutions of the wave equation: particle in a box, rigid rotator, harmonic oscillator, and hydrogenic atom. In each case the allowed solutions were discrete in nature. One finding is that the solutions are a function of integer numbers, known as quantum numbers. In addition, for some cases, multiple allowed solutions resulted in the atom or molecule having the same energy. This is known as degeneracy.

The important results for thermodynamics are the quantum numbers, the allowed energies, and the degeneracy associated with each energy. These are summarized in Table 4.5.

For real atoms and molecules, the situation is more complex. For atoms, the energies and degeneracies do not follow the simple rules of the hydrogenic atom. One must take into account angular momentum and the coupling of various rotational motions. These lead to quite different results, and are difficult to calculate without advanced numerical methods. Furthermore, the Pauli exclusion principle must be taken into account. We used this effect to build up the periodic table. Typically, for energies and degeneracies we utilize data compilations such as are available in the NIST Chemistry Webbook. This is also true of molecules, although there are empirical algebraic expressions for diatomic molecules that can be used to calculate properties. However, for complex situations we are forced to use data from experiments and/or advanced numerical calculations.

4.8 Problems

4.1 The umbrella of ozone in the upper atmosphere is formed from the photolysis of O_2 molecules by solar radiation according to the reaction

$$O_2 + h\nu \rightarrow O + O$$

Calculate the cutoff wavelength above which this reaction cannot occur. Then find a plot of the solar spectrum in the upper atmosphere and identify which portion of the spectrum will cause the O_2 to dissociate.

4.2 What is the physical interpretation of the wave function? How is this quantity obtained and once known, how may it be used?

4.3 Why is the wave function normalized?

4.4 What is meant by the "expectation value" of a dynamical variable?

4.5 What is degeneracy?

4.6 What is the expectation value of the x-momentum of a particle in a box of dimension L on each side?

4.7 What assumption enabled the description of internal motion in spherical coordinates?

4.8 For a molecule that can be modeled as a rigid rotator, what is the degeneracy of a rotational state with quantum number $J = 23$?

4.9 What is the physical significance of m_l?

4.10 Consider an O_2 molecule. Assume that the x-direction translational energy, the rotational energy, and the vibrational energy each equal about $kT/2$. What would the translational, rotational, and vibrational quantum numbers be at 300 K and a volume of 1 cm^3?

4.11 A combustion engineer suggests that a practical means of eliminating NO from engine exhaust is by photodecomposition according to the reaction

$$\mathrm{NO} + h\nu \rightarrow \mathrm{N+O}$$

He suggests using He–Ne laser beams at 632.8 nm as the radiation source. Would the idea work?

4.12 For nitrogen molecules occupying a volume of 1 cm^3, calculate the translational energies corresponding to states for which (n_1, n_2, n_3) are (1, 1, 1), (100, 100, 100), and (1000, 1000, 1000).

4.13 Consider the energy levels of a hydrogenic atom for which $n = 3$. In terms of appropriate quantum number values, list the different eigenstates, first assuming no spin–orbit coupling and then with spin–orbit coupling. Compare the degeneracies for the two cases (including electron spin) and sketch the component levels that result from spin–orbit coupling. Use Fig. 4.13 as a guide.

4.14 For the diatomic molecule assigned to you in class (one per person), calculate and plot the ground electronic state vibrational and rotational-level energies. The plot should be drawn like the figure below, but also include rotational levels. Include the first five vibrational levels and the first five rotational levels for each v. Remember you can get the harmonic oscillator constant k from ω_e.

4.15 For the diatomic molecule I_2, determine the rotational quantum number for the rotational energy state that is equal in value to the energy difference between two adjacent vibrational energy levels. Assume the rigid rotator/harmonic oscillator solution applies.

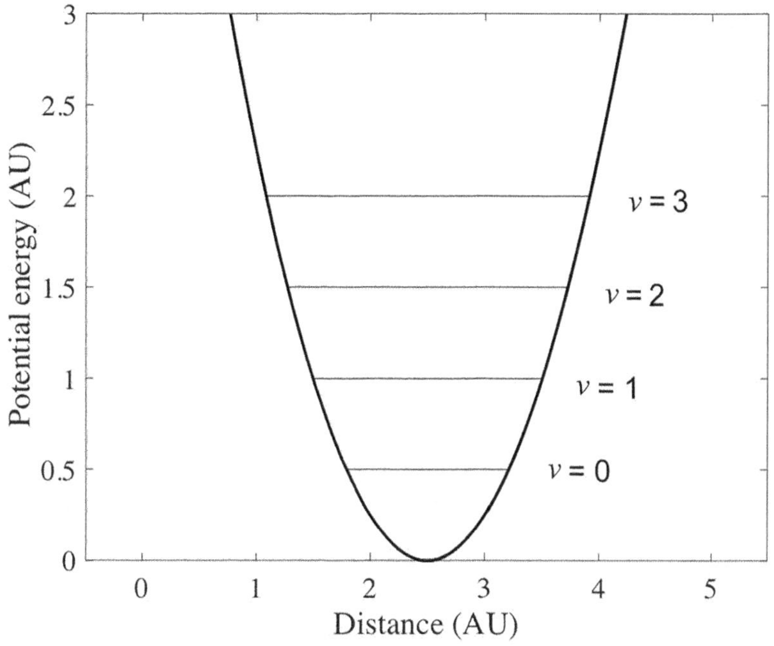

Energy levels of diatomic molecule.

4.16 Carry out a series of calculations on the same diatomic molecule you were assigned in class. The sequence of steps is given below. Do this for two semi-empirical methods (AM1 and PM3) and then two *ab initio* methods. For the two *ab initio* methods choose the Small(3-21G) and Large(6-31G**) basis sets. In all cases just use the other default settings.

(a) Draw the molecule.
(b) Select a method (semi-empirical or *ab initio*).
(c) Optimize the geometry and note the bond length.
(d) Do a single point calculation.
(e) Do a vibrational calculation.
(f) Display the vibrational results and note the vibrational frequency.

Use the NIST database to obtain experimental values of the bond length and vibrational frequency. Compare these with your calculated results. Turn in a summary of your findings.

5 Ideal Gases

Here we deal with non-interacting, or ideal gases. We find that we are in a position to calculate the partition function for such gases in a fairly straightforward way. We shall see that the partition function can be separated into several components, representing each of the various modes of motion available to the atom or molecule. In some cases, simple algebraic relations are obtained. For the monatomic gas, only translational and electronic motion must be considered. For molecules, vibrational and rotational motion must be added. This task is simplified if the Maxwell–Botzmann limit holds, and we will make that assumption for monatomic and molecular gases.

5.1 The Partition Function

We start with the expression for the single-component grand canonical partition function in a gas that satisfies the Maxwell–Boltzmann limit:

$$\ln Q_{MB} = \sum_k e^{-\beta\epsilon_k-\gamma} \tag{5.1}$$

Because γ is not a function of the eigen-index k, we can factor $e^{-\gamma}$ out of the summation so that

$$\ln Q_{MB} = e^{-\gamma}\sum_k e^{-\beta\epsilon_k} \tag{5.2}$$

The sum $q = \sum_k e^{-\beta\epsilon_k}$ is called the "molecular partition function." We can separate the molecular partition function by noting that the energy can be written as the sum of the translational and internal energy

$$\epsilon = \epsilon_{tr} + \epsilon_{int} \tag{5.3}$$

and that in general

$$e^{A+B} = e^A e^B \tag{5.4}$$

If A and B are statistically independent of each other, then

$$\sum_{A,B} e^{-A}e^{-B} = \sum_A e^{-A}\sum_B e^{-B} \tag{5.5}$$

Thus, we can write

$$\ln Q_{MB} = e^{-\gamma} \sum_k e^{-\beta\epsilon_{tr,k}} \sum_k e^{-\beta\epsilon_{int,k}} \tag{5.6}$$

and

$$\ln Q_{MB} = e^{-\gamma} q_{tr} q_{int} \tag{5.7}$$

where the q's are the molecular partition functions. It remains to evaluate them.

5.2 The Translational Partition Function

To evaluate the translational partition function requires the quantum-mechanical expression for the translational energy, which is

$$\epsilon_{tr} = \frac{h^2}{8m}\left\{\left(\frac{n_x}{l_x}\right)^2 + \left(\frac{n_y}{l_y}\right)^2 + \left(\frac{n_z}{l_z}\right)^2\right\} \tag{5.8}$$

Here, the n's are the translational quantum numbers and the l's the macroscopic dimensions in the three coordinate directions. Therefore, if the motions in the three coordinate directions are statistically independent, then

$$q_{tr} = q_x q_y q_z = \sum_{n_x} e^{-\frac{\beta h^2}{8m}\frac{n_x{}^2}{l_x{}^2}} \sum_{n_y} e^{-\frac{\beta h^2}{8m}\frac{n_y{}^2}{l_y{}^2}} \sum_{n_z} e^{-\frac{\beta h^2}{8m}\frac{n_z{}^2}{l_z{}^2}} \tag{5.9}$$

We can obtain an analytic expression for q_{tr} if we note that, in practice, we are dealing with very large quantum numbers. Thus, we can treat the exponentials as continuous rather than discrete functions. For example,

$$q_x = \sum_{n_x} e^{-\frac{\beta h^2}{8m}\frac{n_x{}^2}{l_x{}^2}} \cong \int_0^\infty e^{-\frac{\beta h^2}{8m}\frac{n_x{}^2}{l_x{}^2}} dn_x \tag{5.10}$$

If we let $x = \left(\frac{\beta h^2}{8m}\right)^{1/2} \frac{n_x}{l_x}$, then

$$q_x \cong \left(\frac{8m}{\beta h^2}\right)^{1/2} l_x \int_0^\infty e^{-x^2} dx \tag{5.11}$$

The value of the definite integral can be found in math tables and is equal to $\sqrt{\pi}/2$. Substituting that value and repeating for each coordinate direction, we finally obtain ($\beta = 1/kT$)

$$q_{tr} = \left(\frac{2\pi mkT}{h^2}\right)^{1/2} V \tag{5.12}$$

where $V = l_x l_y l_z$.

Let us now explore the distribution of particles over the allowed translational energy states. Recall that the number of particles in a given state is

$$N_k = e^{-\gamma} e^{-\beta\epsilon_k} \tag{5.13}$$

and since

$$N = e^{-\gamma} \sum_k e^{-\beta \epsilon_k} \tag{5.14}$$

the probability of finding a particle in a given state is

$$\frac{N_k}{N} = \frac{e^{-\beta \epsilon_k}}{\sum_k e^{-\beta \epsilon_k}} \tag{5.15}$$

This is called the Boltzmann ratio and is generally applicable for any mode of motion. For translational energy the differences between energy levels are so small that energy varies continuously, and we can write

$$f(\epsilon) d\epsilon = \frac{e^{-\beta \epsilon_k}}{\sum_k e^{-\beta \epsilon_k}} dg \tag{5.16}$$

where f is a probability distribution function, $f(\epsilon)d\epsilon$ is the fraction of particles in the interval ϵ to $\epsilon + d\epsilon$, and dg is the degeneracy for the energy increment. The denominator is, of course, the translational energy partition function that we have already derived. To relate $d\epsilon$ to dg first consider that

$$\epsilon = \frac{h^2}{8mV^{2/3}} n^2 \tag{5.17}$$

where n is the total translational quantum number equal to $\sqrt{n_x^2 + n_y^2 + n_z^2}$. Therefore, n is the radius of a one-eighth sphere in the positive octant of translational quantum number space. All states on the surface of the sphere have the same energy. Therefore, each state with energy between $\epsilon + d\epsilon$ occupies the one-eighth spherical shell between n and $n + dn$. The volume of this shell is

$$dg = \frac{4\pi n^2 dn}{8} \tag{5.18}$$

Differentiating ϵ with respect to n and substituting into Eq. (5.18), one obtains

$$dg = 2\pi V \left(\frac{2m}{h^2} \right) \sqrt{\epsilon} d\epsilon \tag{5.19}$$

Substituting this into Eq. (5.16), we obtain

$$f(\epsilon) = \frac{2\epsilon^{1/2}}{\pi^{1/2}(kT)^{3/2}} e^{-\epsilon/kT} \tag{5.20}$$

Now, the translational energy of an individual molecule is

$$\epsilon = \frac{mc^2}{2} \tag{5.21}$$

where c is the particle speed, so that

$$d\epsilon = mcdc \tag{5.22}$$

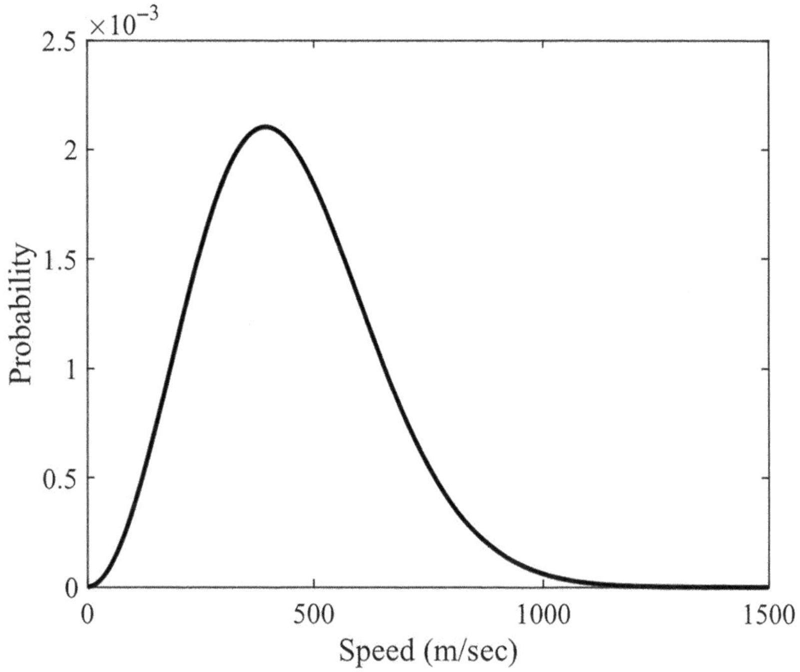

Figure 5.1 The Maxwellian speed distribution.

Note that we must have

$$f(\epsilon)d\epsilon = f(c)dc \tag{5.23}$$

which follows from the idea that the probability of a particle assuming a certain energy must correspond to a certain speed. Substituting Eqs (5.21)–(5.23) into Eq. (5.20), we obtain

$$f(c) = \left(\frac{2}{\pi}\right)^{1/2}\left(\frac{m}{kT}\right)^{3/2} c^2 e^{-mc^2/2kT} \tag{5.24}$$

This relationship, which plays a very important role in the kinetic theory of gases, is called the Maxwellian speed distribution and is illustrated in Fig. 5.1.

Useful properties that can be obtained from the Maxwellian speed distribution include the average speed

$$\bar{c} = \int_0^\infty cf(c)dc = \left(\frac{8kT}{\pi m}\right)^{1/2} \tag{5.25}$$

the most probable speed (obtained by differentiating the Maxwellian and setting it to zero)

$$c_{mp} = \left(\frac{2kT}{m}\right) \tag{5.26}$$

and the root-mean-squared speed

$$\sqrt{c^2} = \int_0^\infty c^2 f(c)dc = \frac{3}{2}\left(\frac{2kT}{m}\right) = \frac{3}{2}c_{mp} \tag{5.27}$$

5.3 Monatomic Gases

Let us next consider a gas made up entirely of atoms. In that case the only other form of motion is that of the electrons orbiting the nucleus. The electronic partition function

$$q_e = \sum_k e^{-\beta\epsilon_k} \tag{5.28}$$

is a sum over quantum states. However, electronic energy states, in contrast to **quantum** states, are degenerate. Thus, we could also write

$$q_e = \sum_k g_k e^{-\beta\epsilon_k} = g_0 + g_1 e^{-\frac{\epsilon_1}{kT}} + g_2 e^{-\frac{\epsilon_2}{kT}} + \cdots \tag{5.29}$$

where now the summation is over energy states rather than quantum states.

To evaluate q_e we need to know the details for each specific atom. The electronic states of many atoms can be found in the NIST Atomic Spectra Database (https://www.nist.gov/pml/atomic-spectra-database). However, recall that electronic energies are quite large, typically 20,000–80,000 K in terms of characteristic temperature. Therefore, at reasonable temperatures, only a few terms in the summation are significant. In fact, for most atoms, the room-temperature partition function is essentially equal to the first term in the summation. If that is the case, then

$$q_e = g_0 \tag{5.30}$$

5.3.1 Example 5.1

Consider the lithium atom, Li. Calculate q_e as a function of temperature. Determine the temperature at which the partition function is increased by 10% above the case where there is no excitation.

The first few electronic states of Li are given in the table below:

Optical electron configuration	Term classification	Energy (cm^{-1})	Degeneracy
$1s^2 2s$	$^2S_{1/2}$	0	2
$1s^2 2p$	$^2P_{1/2}$	14,903.622	2
$1s^2 2p$	$^2P_{3/2}$	14,903.957	4
$1s^2 3s$	$^2S_{1/2}$	27,206.066	2

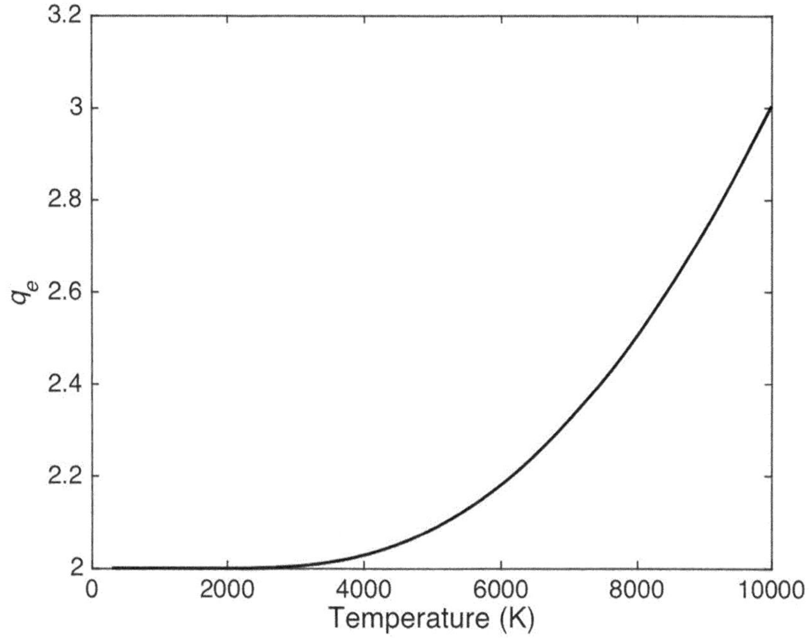

Figure 5.2 Electronic partition function for Li.

q_e is calculated using Eq. (5.29) and plotted in Fig. 5.2. The temperature at which q_e increases by 10% is 6151 K. This illustrates why, at temperatures not exceeding 2000–3000 K, electronic excitation can be ignored.

Ignoring electronic excitation, the fundamental relation for a monatomic ideal gas with no electronic excitation becomes

$$S\left[\frac{1}{T},\frac{-\mu}{T}\right] = ke^{-\frac{\mu}{kT}}\left(\frac{2\pi mkT}{h^2}\right)^{3/2} Vg_e \tag{5.31}$$

Now, recall that

$$dS\left[\frac{1}{T},\frac{-\mu}{T}\right] = -Ud\frac{1}{T} + \frac{p}{T}dV + Nd\frac{\mu}{T} \tag{5.32}$$

where $-U, p/T$, and N are the equations of state in the grand canonical representation. Evaluating N,

$$N = \frac{\partial S\left[\frac{1}{T},\frac{-\mu}{T}\right]}{\partial\frac{-\mu}{T}} = \frac{1}{k}S\left[\frac{1}{T},\frac{-\mu}{T}\right] = \frac{pV}{kT} \tag{5.33}$$

This, of course, is the perfect gas law.

Evaluating U, we obtain

$$U = \frac{-\partial S\left[\frac{1}{T},\frac{-\mu}{T}\right]}{\partial\frac{1}{T}} = \frac{3}{2}TS\left[\frac{1}{T},\frac{-\mu}{T}\right] = \frac{3}{2}pV = \frac{3}{2}NkT \tag{5.34}$$

The average internal energy per atom is

$$u = \frac{U}{N} = \frac{3}{2}kT \tag{5.35}$$

The entropy becomes

$$S = S\left[\frac{1}{T}, \frac{-\mu}{T}\right] + U\frac{1}{T} - N\frac{\mu}{T} = k\left[N + U\frac{1}{T} - N\frac{\mu}{T}\right] = k\left[\frac{5}{2} - \frac{\mu}{kT}\right] \tag{5.36}$$

Finally, it is useful to calculate the specific heat c_v,

$$c_v = \left.\frac{\partial u}{\partial T}\right)_V = \frac{3}{2}k \tag{5.37}$$

Each of these property expressions is valid if our assumption about the electronic partition function holds. If not, then q_e is a function of temperature, and the derivatives must be evaluated taking that into account. We can assess the importance of higher-order electronic terms using the Boltzmann ratio. It can be rewritten as the ratio of the populations of two states:

$$\frac{N_k}{N_j} = \frac{g_k e^{-\beta\epsilon_k}}{g_j e^{-\beta\epsilon_j}} \tag{5.38}$$

For example, consider the ratio of the first excited electronic state of the nitrogen atom to the ground state. For nitrogen, ϵ_1 = 19,227 cm^{-1} and g_0, g_1 = 4, 10, respectively. If the temperature is 300 K, then

$$\frac{N_1}{N_0} = \frac{10e^{-\frac{19227hc}{kT}}}{4} = 1.67 \times 10^{-40} \tag{5.39}$$

However, if the temperature were 2000 K, then the ratio would be about 2.4×10^{-6}. Clearly, the Boltzmann ratio is a very strong function of temperature. Even so, for nitrogen only the first term of the electronic partition function sum is important at these temperatures. For other atomic species, this may not be the case, and one must treat each specie on a case-by-case basis.

Note, finally, that we can fairly precisely evaluate the limits of the Maxwell–Boltzmann approximation. The exact criterion is that

$$e^{\beta\epsilon_k + \gamma} >> 1 \tag{5.40}$$

for all values of k. Since $e^{\beta\epsilon_k} \geq 1$ for all values of k, the criterion must apply to e^{γ}. Using the fundamental relation to calculate e^{γ}, we obtain

$$e^{\gamma} = \left(\frac{2\pi mkT}{h^2}\right)^{3/2} \frac{kT}{p} g_e \tag{5.41}$$

Since the electronic degeneracy is order unity, the inequality can be written

$$\left(\frac{2\pi mkT}{h^2}\right)^{3/2} \frac{kT}{p} g_e >> 1 \tag{5.42}$$

For hydrogen at 300 K and 1 atm, the left-hand side has a value of 3.9×10^{34}, clearly satisfying the inequality.

5.4 Diatomic Gases

For a diatomic gas, the total energy is

$$\epsilon = \epsilon_{tr} + \epsilon_e + \epsilon_v + \epsilon_r \tag{5.43}$$

Following the arguments made in discussing the monatomic gas, we write the internal energy portion of the partition function as

$$\sum_k e^{-\beta\epsilon_k} = \sum_k e^{-\beta\epsilon_{ktr}} \sum_k e^{-\beta\epsilon_{ke}} \sum_k e^{-\beta\epsilon_{kv}} \sum_k e^{-\beta\epsilon_{kr}} = q_{tr}q_e q_v q_r \tag{5.44}$$

Thus, the partition function becomes

$$\ln Q = e^{-\gamma} q_{tr} q_e q_v q_r \tag{5.45}$$

We have already discussed the evaluation of the translational and electronic partition functions. We now consider rotation and vibration.

5.4.1 Rotation

Recall that for a rigid rotator,

$$\epsilon_r = k\Theta_r J(J+1) \tag{5.46}$$

where Θ_r is the characteristic rotational temperature. The rotational degeneracy is $g_J = 2J + 1$. Thus,

$$q_r = \sum_J (2J+1)e^{-\frac{\epsilon_J}{kT}} = \sum_J (2J+1)e^{-\frac{\Theta_r}{T}J(J+1)} \tag{5.47}$$

One can approach the evaluation of this sum in several ways. The most accurate is to directly evaluate it numerically. The number of terms required will depend on the ratio Θ/T. The higher the temperature, the more terms will be required. Alternatively, for $(\Theta/T)J(J+1) << 1$, the sum can be approximated by an integral

$$q_R \simeq \int_{J=0}^{\infty} (2J+1)e^{-\frac{\Theta_R}{T}J(J+1)} dJ \tag{5.48}$$

Defining $u = J(J+1)$, the integral becomes

$$\int_0^{\infty} e^{-\frac{\Theta_R}{T}u} du = \frac{T}{\theta_R} \tag{5.49}$$

This expression is referred to as the high-temperature limit. As shown in Hirschfelder, Curtiss, and Bird [14] and McQuarrie [31], the integral approximation is really the first term in an Euler–MacLaurin summation:

$$\sum_{n=a}^{b} f(n) = \int_a^b f(n)dn + \frac{1}{2}\{f(b) + f(a)\}$$
$$+ \sum_{j=1}^{\infty} (-)^j \{f^{(2j-1)}(a) - f^{(2j-1)}(b)\} \quad (5.50)$$

The full expression becomes

$$q_r = \frac{T}{\Theta_r}\left[1 + \frac{1}{3}\frac{\Theta_r}{T} + \frac{1}{15}\left(\frac{\Theta_r}{T}\right)^2 + \frac{4}{315}\left(\frac{\Theta_r}{T}\right)^3 + \cdots\right] \quad (5.51)$$

There is a great advantage in using Eq. (5.51) to calculate the partition function at lower temperatures in that it converges quite rapidly compared to evaluating the sum for q_R directly.

There is a complication when the molecule is homonuclear (i.e. both atoms are the same type). For homonuclear molecules such as H_2, O_2, N_2, and so on, symmetry requirements prevent all possible values of the rotational quantum number. In fact, only even or odd ones are allowed, so that generally

$$q_r = \frac{1}{\sigma}\frac{T}{\Theta_r} \quad (5.52)$$

where σ is a symmetry factor and

$\sigma = 1$ for heteronuclear
$\sigma = 2$ for homonuclear

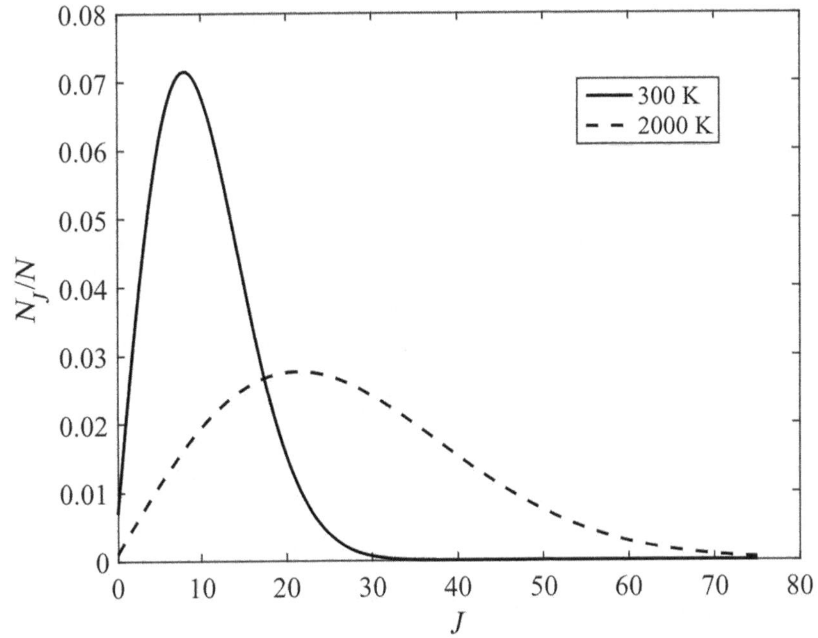

Figure 5.3 Equilibrium rotational population distribution for CO.

It is useful to explore the distribution of rotational states. This is given by the Boltzmann ratio

$$\frac{N_J}{N} = \frac{g_J e^{-J(J+1)\frac{\Theta_r}{T}}}{q_r(T)} \tag{5.53}$$

This expression is plotted in Fig. 5.3 for CO at 300 K and 2000 K. The peak of the curve can be found by differentiating N_J/N. The result is

$$J_{\max} \approx \left(\frac{T}{2\Theta_r}\right)^{1/2} \tag{5.54}$$

As the temperature increases, the distribution broadens as more and more energy levels are populated.

5.4.2 Example 5.2

Calculate and plot q_r as a function of the number of terms in the sum for the molecule OH at 300, 1000, and 2000 K. For each temperature, also include in the plot the integral result Θ/T. Also, plot the percentage error involved in using the simple high-temperature expression.

This is easily done in Matlab or Mathematica. See Figs 5.4 and 5.5.

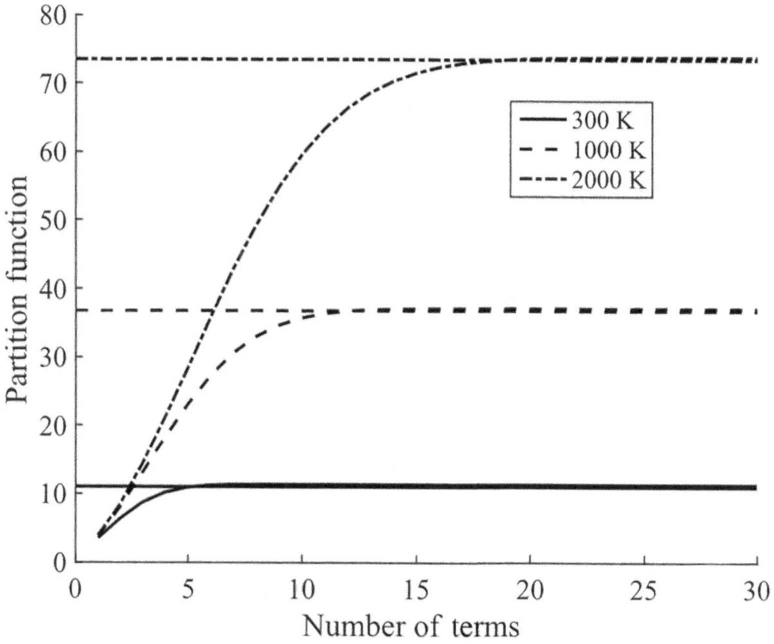

Figure 5.4 Effect of number of terms on the value of q_r. Horizontal lines are the high-temperature limit. Although not plotted, the Euler–MacLaurin series converges in only a few terms.

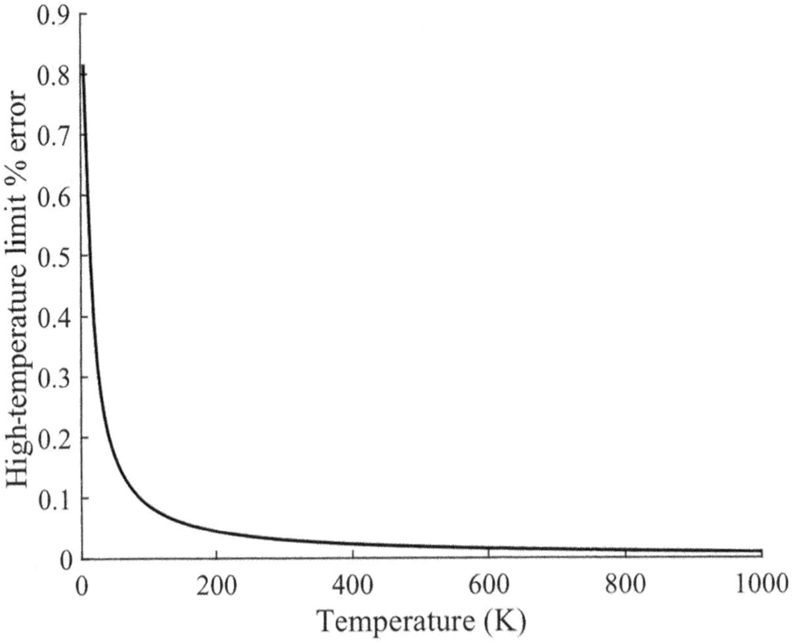

Figure 5.5 Effect of number of terms on error using the high-temperature limit.

5.4.3 Vibration

The vibrational energy for a harmonic oscillator is

$$\epsilon_v = h\nu_v(v + 1/2) \tag{5.55}$$

However, before we use this expression to calculate the vibrational partition function, an issue of energy referencing must be resolved. The potential function of a real diatomic molecule will look similar to that shown in Fig. 5.6. In this coordinate system, the absolute energy of vibration is set to zero when there is a complete separation of the atoms that make up the molecule. This makes sense because

$$P.E. = -\int_r^{\infty} F(r)dr \tag{5.56}$$

and as the atoms approach from a distance, there is initially no force and the potential energy is zero. As the distance closes, the electrostatic interaction force is initially attractive and the potential energy decreases. When the force becomes repulsive, the potential energy increases. At small r it reaches very large values.

Because of this, and because referencing is important when working with reacting mixtures, it is usual to set the same zero reference when calculating the partition function. Therefore, we will write the vibrational energy as

$$\epsilon_v = h\nu_v(v + 1/2) - D_e \tag{5.57}$$

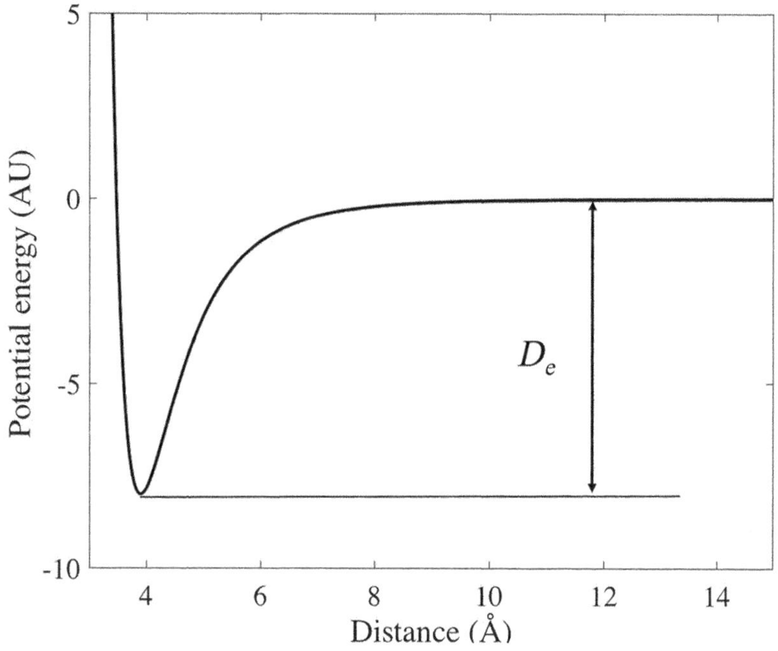

Figure 5.6 Typical potential energy function.

Here D_e is called the electronic binding energy, and is the total depth of the potential well. (Note, this is different from the dissociation energy, D_0, which is D_e minus the zero-point energy and is the energy required to dissociate the molecule from the ground vibrational state.)

The partition function thus becomes

$$q_v = e^{\frac{D_e}{kT}} \sum_v e^{-\frac{h\nu(v+1/2)}{kT}} = e^{\frac{D_e}{kT}} \sum_v e^{-\frac{\Theta_v}{T}(v+1/2)} \tag{5.58}$$

The sum can be expanded as

$$\sum_v e^{-\frac{\Theta_v}{T}(v+1/2)} = \sum_v e^{-\frac{\Theta_v}{2T}(2v+1)} = e^{-\frac{\Theta_v}{2T}}\left[1 + e^{-\frac{\Theta_v}{T}} + e^{-2\frac{\Theta_v}{T}} + \cdots\right] \tag{5.59}$$

The term in brackets is a binomial expansion. Thus, the partition function can be written as

$$q_v = e^{\frac{D_e}{kT}} \frac{e^{-\frac{\Theta_v}{2T}}}{1 - e^{-\frac{\Theta_v}{T}}} = e^{\frac{D_e}{kT}} \frac{1}{2\sinh\left(\Theta_v/2T\right)} \tag{5.60}$$

5.4.4 Properties

Substituting all the molecular partition functions into the full partition function, we have the fundamental relation for $T >> \Theta_r$ and $q_e = g_e$:

$$S\left[\frac{1}{T}, -\frac{\mu}{T}\right] = e^{-\frac{\mu}{kT}} \left(\frac{2\pi mkT}{h^2}\right)^{3/2} V \left(\frac{T}{\sigma\Theta_r}\right) \frac{e^{\frac{D_e}{kT}}}{2\sinh(\Theta_v/2T)} g_e \tag{5.61}$$

If we evaluate the equations of state, we would find that the perfect gas law still holds, and that the energy per molecule is

$$u = \frac{3}{2}kT + kT + \left[-D_e + \frac{k\Theta_v}{2}\coth\frac{\Theta_v}{T}\right] \tag{5.62}$$

The other properties of interest are the entropy

$$s = \frac{S}{N} = k\left[\frac{7}{2} + \left(\frac{\Theta_v}{2T}\coth\frac{\Theta_v}{2T} - D_e\right) - \frac{\mu}{kT}\right] \tag{5.63}$$

and the specific heat

$$c_v = k\left[\frac{3}{2} + 1 + \left\{\frac{\Theta_v/2T}{\sinh(\Theta_v/2T)}\right\}^2\right] \tag{5.64}$$

The three terms in the expressions for energy and specific heat are due to translation, rotation, and vibration, respectively. For an ideal gas, the translational energy is always $3/2kT$ regardless of the internal structure of the molecule. Thus, the contribution of translation to the specific heat is $3/2k$. Similarly, the contribution due to rotation, when $T >> \Theta_r$, is k. For vibration, however, at temperatures below and near Θ_v, the contribution to the energy, and thus specific heat, is a function of temperature. c_v/k for CO is shown in Fig. 5.7. As can be seen, the curve is S-shaped, representing the fact that at low temperatures all the molecules are in their ground vibrational states and the contribution of vibration is insensitive to temperature. As T increases, more molecules are in higher vibrational states. At high T, the vibrational contribution to the energy

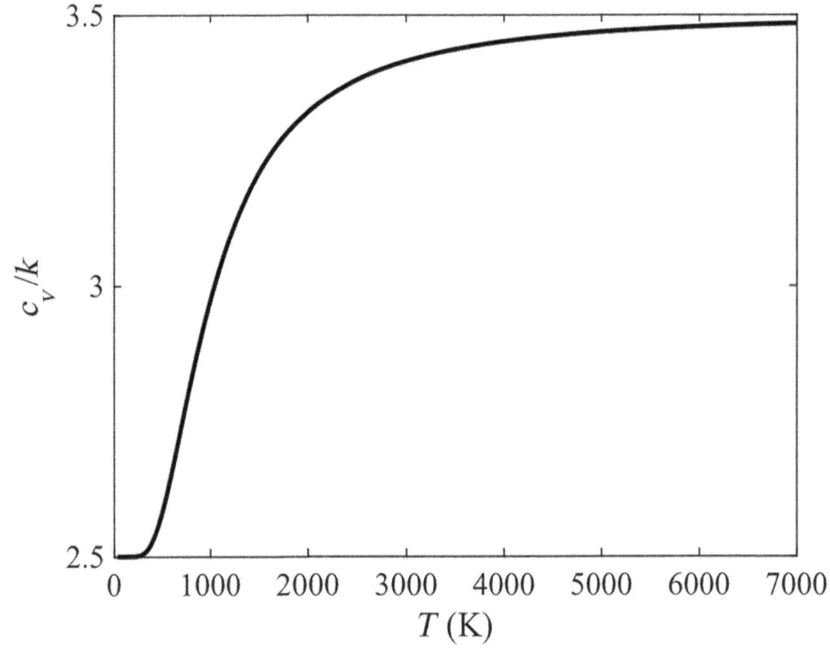

Figure 5.7 c_v/k as a function of temperature for CO.

reaches the limiting value of kT. Therefore, as T increases from values of $T << \Theta_v$ to values of $T >> \Theta_v$, c_v goes from $5/2k$ to $7/2k$. Once the vibrational contribution to c_v reaches k, vibration is said to be "fully excited." Actually, this phenomenon occurs for all modes of motion. For translation and rotation, full excitation is reached at low temperatures. Electronic motion is never fully excited because dissociation or ionization occurs at lower temperatures than full excitation.

We have seen that real diatomic molecules do not behave exactly like the simple rigid rotator/harmonic oscillator models. If the vibrational and rotational states are merely distorted from the simple model, then one can still separate the molecular partition functions and use the more complete term expressions or experiment data to calculate the q's. However, if there is coupling between the modes, then q cannot be easily separated, and must be directly evaluated. For example, suppose that the sum of vibrational and rotational energies is

$$G(v) + F_v(J) \tag{5.65}$$

where the subscript on F indicates that the rotational energy depends on the vibrational state. Then we must write

$$q_{v,r} = \frac{1}{\sigma}\sum_v \sum_J (2J+1)e^{-(hc/kT)[G(v)+F_v(J)]} \tag{5.66}$$

In practice, one must examine each case to determine whether the simple models provide sufficient accuracy for a given application. Most property compilations have used numerical, rather than analytic procedures to calculate the partition functions and the equations of state. (Details regarding numerical procedures can be found in the NIST-JANNAF Themochemical Tables [32].)

5.5 Polyatomic Gases

The principles for evaluating the partition function for polyatomic gases are similar to those we have already applied. The results for translational energy are identical to those for monatomic and diatomic gas assemblies, as translational energy describes the motion of the center of mass of any molecular structure. However, evaluating the internal modes of energy is somewhat more difficult because of the many modes of motion and the likelihood that they interact.

For a body made up of n atoms, there are $3n$ degrees of freedom. Three of these are taken up by translational motion, leaving $3n - 3$ modes for vibrational and rotational motion. For a linear molecule, such as CO_2, there are only two rotational modes, thus there are $3n - 5$ possible vibrational modes. For the general nonlinear case, there are three rotational modes and thus $3n - 6$ vibrational modes.

If we assume that each vibrational and rotational mode is independent, then we can separate the partition function for each mode. Furthermore, if we assume that the harmonic oscillator and rigid rotator models apply, then we can write for the general case

$$q_v = e^{\frac{D_e}{kT}} \prod_i^{3n-6} \left\{ \frac{e^{-\Theta_{v_i}/2T}}{1 - e^{-\Theta_{v_i}/T}} \right\} \tag{5.67}$$

and

$$q_r = \frac{1}{\sigma} \left\{ \frac{\pi T^3}{\Theta_{r_1}\Theta_{r_2}\Theta_{r_3}} \right\} \tag{5.68}$$

where σ is the rotational symmetry factor. As with atoms and diatomic molecules, however, real polyatomic behavior rarely follows the simple models and numerical methods must be used to evaluate the partition function.

There is a concept called the "equipartition of energy." We have noted that "fully excited" translational energy contributes $\frac{3}{2}kT$ to c_v, rotation kT for each axis of rotation, and kT each for the potential and kinetic energy of each vibrational mode. This concept can be used to estimate c_v in large molecules, as illustrated by Problem 5.7.

5.6 Summary

In this chapter we began the process of relating macroscopic thermodynamic properties to the partition function for specific situations, non-interacting gases in this case. These include ideal gases, reacting mixtures of ideal gases, the photon gas, and the electron gas.

Our first development involved defining the molecular partition functions. Recall that the grand canonical partition function for a gas that satisfies the Maxwell–Boltzmann limit is

$$\ln Q_{MB} = \sum_k e^{-\beta\epsilon_k - \gamma} \tag{5.69}$$

Factoring $e^{-\gamma}$ out of the sum we obtain

$$\ln Q_{MB} = e^{-\gamma} \sum_k e^{-\beta\epsilon_k} \tag{5.70}$$

The sum $q = \sum_k e^{-\beta\epsilon_k}$ is called the molecular partition function, and because of the independence of the various modes of motion we can write (as appropriate)

$$\ln Q_{MB} = e^{-\gamma} q_{tr} q_{rot} q_{vib} q_e \tag{5.71}$$

We then evaluated each of these molecular partition functions.

5.6.1 Monatomic Gas

For monatomic gases, the only modes of motion are translation and electronic. Starting with translation, we obtained

$$q_{tr} = \left(\frac{2\pi mkT}{h^2} \right)^{1/2} V \tag{5.72}$$

While we were able to derive this algebraic expression for translation, the electronic partition function must be calculated numerically. In many cases, the first excited electronic states are so energetic that all but the first term in the partition function can be ignored. In that case, $q_e = g_e$. Thus, combining with the translational partition function we obtain

$$\ln Q = e^{-\gamma}\left(\frac{2\pi mkT}{h^2}\right)^{1/2} V g_e \tag{5.73}$$

Thus, the fundamental relation for a monatomic ideal gas with no electronic excitation becomes

$$S\left[\frac{1}{T}, \frac{-\mu}{T}\right] = k e^{-\frac{\mu}{kT}}\left(\frac{2\pi mkT}{h^2}\right)^{3/2} V g_e \tag{5.74}$$

5.6.2 Simple Diatomic Gas

For a diatomic gas we also took rotation and vibration into account. The molecular partition functions for these are

$$q_r = \frac{1}{\sigma}\frac{T}{\Theta_r} \tag{5.75}$$

where σ is a symmetry factor and

$$\begin{aligned} \sigma &= 1 \quad \text{for heteronuclear} \\ \sigma &= 2 \quad \text{for homonuclear} \end{aligned}$$

$$q_v = e^{\frac{D_e}{kT}} \frac{e^{-\frac{\Theta_v}{2T}}}{1 - e^{-\frac{\Theta_v}{T}}} = e^{\frac{D_e}{kT}} \frac{1}{2\sinh\left(\Theta_v/2T\right)} \tag{5.76}$$

Substituting all the molecular partition functions into the full partition function, we have the fundamental relation for $T >> \Theta_r$ and $q_e = g_e$:

$$S\left[\frac{1}{T}, -\frac{\mu}{T}\right] = e^{-\frac{\mu}{kT}}\left(\frac{2\pi mkT}{h^2}\right)^{3/2} V \left(\frac{T}{\sigma\Theta_r}\right) \frac{e^{\frac{D_e}{kT}}}{2\sinh(\Theta_v/2T)} g_e \tag{5.77}$$

Of course, this expression is only valid for the rigid rotator/harmonic oscillator. As we discussed in this chapter, real diatomic molecules are more complex.

5.6.3 Polyatomic Molecules

Recall that for a body made up of n atoms, there are $3n$ degrees of freedom. Three of these are taken up by translational motion, leaving $3n - 3$ modes for vibrational and rotational motion. For a linear molecule, such as CO_2, there are only two rotational modes, thus there are $3n - 5$ possible vibrational modes. For the general nonlinear case, there are three rotational modes and thus $3n - 6$ vibrational modes.

If we assume that the rigid rotator approximation holds and that the various modes of vibration in a polyatomic molecule are independent, then the simple expressions for the rotational and vibrational partitions given in Section 5.5 hold:

$$q_v = e^{\frac{D_e}{kT}} \prod_i^{3n-6} \left\{ \frac{e^{-\Theta_{v_i}/2T}}{1 - e^{-\Theta_{v_i}/T}} \right\} \tag{5.78}$$

and

$$q_r = \frac{1}{\sigma}\left\{\frac{\pi T^3}{\Theta_{r_1}\Theta_{r_2}\Theta_{r_3}}\right\} \tag{5.79}$$

Be sure to note that for a linear molecule, CO_2 for example, there are only two degrees of rotational freedom.

5.7 Problems

5.1 Calculate the value of the translational partition function of O_2 at 300 and 1000 K for a volume of 1 m^3.

5.2 Calculate the average molecular speed of O_2 at 300 and 1000 K.

5.3 For H_2, find the temperature at which Eq. (5.42) is satisfied such that the left-hand side is equal to 10^6.

5.4 Plot the population ratio of electronic state 1 to state 0 for sodium atoms from 300 to 3000 K. Will electronic excitation contribute significantly to the electronic partition function anywhere within this range?

5.5 Plot the rotational partition function of N_2 as a function of temperature from 10 to 300 K. At what temperature does the approximation of Eq. (5.49) result in less than a 1% error?

5.6 Plot the vibrational partition function for CO from 300 to 1000 K. Ignore the dissociation energy term.

5.7 Consider the molecule D-glucose, $C_6H_{12}O_6$, the base molecule in cellulose.

(a) Calculate the total number of energy modes and the number of translational, rotational, and vibrational modes. (I want numbers.)

(b) Using the "equipartition of energy" concept, estimate ideal gas c_p (not c_v) in units of k_B for the "low-temperature" regime where no vibrational modes are activated and for the "high-temperature" regime where all the vibrational modes are activated. (I want numbers.)
(*Note:* The hydrogens attached to carbons are not shown. All C–C bonds are single bonds.)

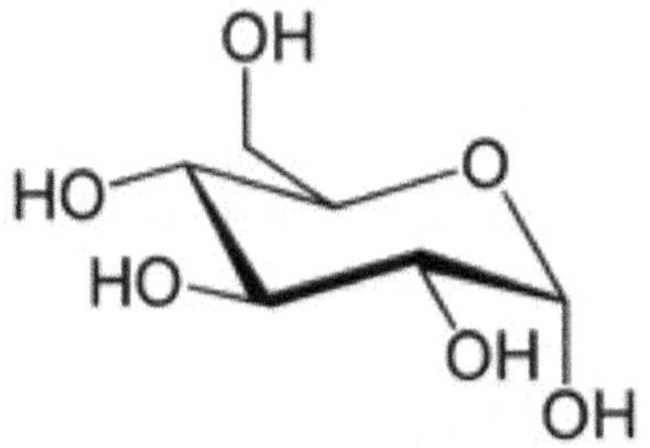

6 Ideal Gas Mixtures

Here we explore the properties of mixtures of ideal gases. For non-reacting gas mixtures of known composition, this is fairly straightforward. For reacting mixtures, a maximization or minimization process is required, depending on the choice of representation.

6.1 Non-reacting Mixtures

The treatment of ideal gas non-reacting mixtures is well covered in most undergraduate textbooks. Here we briefly summarize the important relations. The prediction of the properties of ideal gas mixtures are based on Dalton's law of additive pressures and Amagat's law of additive volumes. Dalton's law states that the pressure of a gas mixture is equal to the sum of the pressures that each component would exert if it were alone at the mixture temperature and volume. Amagat's law states that the volume of a mixture is equal to the sum of the volumes each component would occupy if it were alone at the mixture temperature and pressure. One can derive these results directly by evaluating the partition function for the mixture.

For the Gibbs representation in which temperature, pressure and mass, mole number, or molecular number are the independent variables, the following simple summation relations over all r species thus hold:

$$U = \sum_j^r U_j = \sum_j^r u_j N_j \tag{6.1}$$

$$H = \sum_j^r H_j = \sum_j^r h_j N_j \tag{6.2}$$

$$S = \sum_j^r S_j = \sum_j^r s_j N_j \tag{6.3}$$

for the extensive properties. Similarly for the normalized or specific properties:

$$u = \sum_j^r x_j u_j \tag{6.4}$$

$$h = \sum_j^r x_j h_j \tag{6.5}$$

$$s = \sum_j^r x_j s_j \tag{6.6}$$

$$c_v = \sum_j^r x_j c_{vj} \tag{6.7}$$

where x_j is the mole fraction for specie j, defined as

$$x_j = \frac{N_j}{N_{tot}} \tag{6.8}$$

We have seen that internal energy and enthalpy are a function only of temperature. Thus, in these summations, h_j and u_j are evaluated at the mixture temperature. In the Gibbs representation, however, entropy is a function of both temperature and pressure. The s_j must be evaluated at the mixture temperature and the partial pressure, or

$$s(T,p) = \sum_j^r x_j s_j(T,p_j) \tag{6.9}$$

It is common to rewrite this expression in terms of the mole fraction and total pressure. The Gibbs representation expression for the entropy is

$$s_2 - s_1 = c_p \ln \frac{T_2}{T_1} - R \ln \frac{p_2}{p_1} \tag{6.10}$$

For application to a mixture, this relation is used to relate the entropy at the partial pressure to that at the total pressure. Thus,

$$s_j(T,p_j) = s_j(T,p) + R \ln x_j \tag{6.11}$$

6.1.1 Changes in Properties on Mixing

When a mixture undergoes a state change, calculating the change in properties is straightforward. For example:

$$\Delta u_{mix} = \sum_j^r x_j \Delta u_j \tag{6.12}$$

$$\Delta h_{mix} = \sum_j^r x_j \Delta h_j \tag{6.13}$$

$$\Delta s_{mix} = \sum_j^r x_j \Delta s_j \tag{6.14}$$

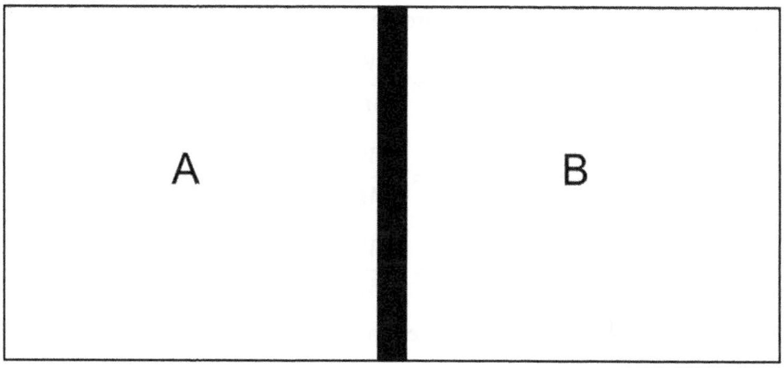

Figure 6.1 Adiabatic mixing at constant total volume.

However, when the ideal gases are mixed, special care must be taken with the entropy because of its pressure dependence:

$$\Delta s_j = s_j(T_2, p_2) - s_j(T_1, p_1) \tag{6.15}$$

Consider the divided volume as shown in Fig. 6.1, where sides A and B of the volume are at the same temperature and pressure, but contain difference species. The barrier between the two sides is then removed and the gases allowed to mix. The change of entropy for the systems becomes

$$\Delta s = -x_A R_A \ln\left(\frac{p_A}{p}\right) - x_B R_B \ln\left(\frac{p_B}{p}\right) \tag{6.16}$$

Since both p_A and p_B are less than p, this means there is an entropy increase because of mixing.

6.1.2 Example 6.1

Methane, ethane, and propane are mixed in a tank. They are supplied separately from storage cylinders at different temperatures (10, 30, and 20°C) and pressures (15, 10, and 6 bar). A pressure relief valve keeps the pressure in the tank at 5 bar. In the final mixture the mole fractions are 0.7, 0.2, and 0.1, respectively. Determine the final temperature in the tank and the total entropy change for the process. (Assume constant specific heats.)

The molecular weights of methane, ethane, and propane are 16.043, 30.070, and 44.097 kg/kmol. c_p = 2.2537, 1.7662, and 1.6794 kJ/kg K. R = 0.5182, 0.2765, and 0.1976 kJ/kg K.

The molar mass of the mixture is

$$\begin{aligned} M_{mix} &= y_{CH_4} MW_{CH_4} + y_{C_2H_6} MW_{C_2H_6} + y_{C_3H_8} MW_{C_3H_8} \\ &= 0.7 \times 16.043 + 0.3 \times 30.070 \times +0.1 \times 44.097 = 21.654 \text{ kg/kmol} \end{aligned}$$

Then, the mass fractions in the mixture become

$$y_{CH_4} = \frac{MW_{CH_4}}{M_{mix}} = 0.519$$

$$y_{C_2H_6} = \frac{MW_{C_2H_6}}{M_{mix}} = 0.278$$

$$y_{C_3H_8} = \frac{MW_{C_3H_*}}{M_{mix}} = 0.204$$

Applying the conservation of energy equation for steady-state flow, one can obtain the temperature of the mixture:

$$y_{CH_4}h_{CH_4}(10°C) + y_{C_2H_6}h_{C_2H_6}(30°C) + y_{C_3H_8}h_{C_3H_8}(20°C)$$
$$= y_{CH_4}h_{CH_4}(T_{mix}) + y_{C_2H_6}h_{C_2H_6}(T_{mix}) + y_{C_3H_8}h_{C_3H_8}(T_{mix})$$

Noting that $\Delta h = c_p(T_2 - T_1)$, the temperature becomes 16.61°C.

The entropy change for each component is

$$\Delta s = c_p \ln\left(\frac{T_2}{T_1}\right) - R\ln\left(\frac{P_2}{P_1}\right)$$

and

$$\Delta s_{mix} = y_{CH_4}\Delta s_{CH_4} + y_{C_2H_6}\Delta s_{C_2H_6} + y_{C_3H_8}\Delta s_{C_3H_8} = 1.011 \text{ kJ/kg K}$$

6.2 Reacting Mixtures

6.2.1 General Case

The most straightforward way to find the equilibrium state of a reacting mixture is to minimize the appropriate thermodynamic potential. For example, if the pressure and temperature are to be controlled, then we could minimize the Gibbs function

$$G(T, p, N_1, \ldots, N_c) \tag{6.17}$$

where r is the total number of compounds being considered. If T and V are held constant, then minimize the Helmholtz function:

$$F(T, V, N_1, \ldots, N_c) \tag{6.18}$$

In differential form these are

$$dG = -SdT + Vdp + \sum_{j}^{r} \mu_j dN_j \tag{6.19}$$

and

$$dF = -SdT - pdV + \sum_{j}^{r} \mu_j dN_j \tag{6.20}$$

where j is the species index. In either case, we hold the independent parameters constant except for the N_j and set the total derivative equal to zero. Thus,

$$\sum_j^r \mu_j dN_j = 0 \tag{6.21}$$

The minimization is subject to a set of constraints, namely that atoms, the basic building blocks of molecules, be conserved and that no mole numbers be negative. We can write

$$b_k = \sum_j^r n_{kj} N_j \tag{6.22}$$

where n_{kj} is the number of k-type atoms in j-type molecules. Thus, b_k is the total number of k-type atoms in the mixture. (The form of the no negative mole number requirement depends on the specific numerical method to be used.)

The temperature, pressure, or volume and number dependency is contained in the equation of state for the μ_j. The ideal gas expression for the chemical potential can be understood by recalling that

$$G \equiv U - TS + pV \tag{6.23}$$

By Euler's relation

$$U = TS - pV + \sum_j^r \mu_j N_j \tag{6.24}$$

Thus,

$$G = \mu_1 N_1 + \mu_2 N_2 + \cdots + \mu_c N_c \tag{6.25}$$

The chemical potential can be thought of as the specific Gibbs function, since

$$g = \frac{G}{N} \tag{6.26}$$

In an ideal gas mixture, specific or partial properties are evaluated at the given temperature and partial pressure of the component in question, or

$$G = \sum_j^r \mu_j(T, p_j) N_j \tag{6.27}$$

However, we need a relationship for the μ_j that is explicit in p_j or x_j. Noting that

$$g = h - Ts = h(T) - Ts(T, p) \tag{6.28}$$

one can write, using Eq. (6.11),

$$\mu_j(T, p_j) = g_j(T, p_j) = g_j(T, p) + RT \ln x_j \tag{6.29}$$

The minimization function thus becomes

$$G = \sum_j^r \left[g_j(T, p) + RT \ln x_j \right] \tag{6.30}$$

There are a number of numerical methods to minimize a function subject to constraints. In Listing 6.1 (see Example 6.2) we use the Matlab function "fmincon" to solve the problem of equilibrium between gaseous water, hydrogen, and oxygen. However, to proceed requires knowing the Gibbs function $g_j(T,p)$.

6.2.2 Properties for Equilibrium and 1st Law Calculations

To use either the general equations of Section 6.2.1 or the equilibrium constants of Section 6.2.4 requires specification of the Gibbs free energy. In addition, if one wants to carry out 1st and 2nd Law analysis, one must be able to calculate other properties such as the internal energy, enthalpy, and entropy. For reacting mixtures it is important to recognize that internal energies must be properly referenced so that energies of reaction, either exothermic or endothermic, are properly accounted for.

The usual way to provide proper referencing is to consider the so-called standard reference state. Every molecule's internal energy, enthalpy, and Gibbs free energy are referenced to the energy required to form the molecule from stable atoms. This is normally done at standard temperature and pressure, and then the relations of Section 6.1 are used to calculate the properties at other temperatures and pressures.

A concerted effort to compile properties was one goal of the JANNAF (Joint Army and Navy NASA Air Force) Interagency Propulsion Committee. JANNAF was formed in 1946 to consolidate rocket propulsion research under one organization. It still functions as the US national resource for worldwide information, data, and analysis on chemical, electrical, and nuclear propulsion for missile, space, and gun propulsion systems. Much of the property work has been carried out by NIST, the National Institute of Standards and Technology. The property values compiled under the JANNAF program were published as the NIST–JANNAF Thermochemical Tables [32]. The tables are available online at http://kinetics.nist.gov/janaf/. The tables provide C_p°, S°, $[G^\circ - H^\circ]/T$, $H - H^\circ(T_r)$, $\Delta_f H^\circ$, $\Delta_f G^\circ$, and log K_f. (The NIST notation for K_p is K_f, where the subscript f stands for formation.)

It is important to understand the definitions of the various parameters in the tables. C_p° is the constant-pressure ideal gas specific heat, as we have discussed before. S° is the absolute entropy evaluated at the standard reference pressure of 1 bar, going to zero at 0 K. The term $H - H^\circ(T_r)$ refers to the change in enthalpy with respect to the standard reference temperature, 298.15 K. Thus, the change in enthalpy between any two temperatures can be obtained as

$$\Delta H = H(T_1 - H^\circ(T_r)) - H(T_2 - H^\circ(T_r)) \tag{6.31}$$

To obtain the entropy at some pressure other than 1 bar, use the relation

$$S(T,p) = s^\circ - R \ln \frac{p}{p_{ref}} \tag{6.32}$$

The JANNAF tables also list the heat of formation $\Delta_f H^\circ$. This is the enthalpy required to form the compound from elements in their standard state at constant temperature and pressure. Thus, oxygen and hydrogen, for example, would be in the diatomic form O_2 and H_2, while carbon would be in the atomic form C. The value of $\Delta_f H^\circ$ is negative when energy is released, as a compound is formed from its elements. To convert $\Delta_f H^\circ$ between the gas and liquid phases, add the heat of vaporization:

$$\Delta_f H^\circ_{298.15}(gas) = \Delta_f H^\circ_{298.15}(liq) + h_{fg,298.15} \tag{6.33}$$

$\Delta_f G^\circ$ is called the Gibbs free energy of formation. It is zero for elements in their equilibrium form, but not for compounds. See Example 6.2 for $\Delta_f G^\circ$ used in a numerical Gibbs free energy minimization. (The term $[G^\circ - H^\circ]/T$ was included in the tables when hand calculations were more common. See the introduction to the JANNAF tables for more information.)

Also listed in the tables are the logarithms of the equilibrium constants, $\log_{10} K_f$, for formation of the compound from elements in their standard state. $\log_{10} K_f$ is zero for elements in their standard state, but nonzero for compounds. We will discuss the equilibrium constants in more detail in Section 6.2.4.

For computational work, it is most convenient to have the properties in the form of algebraic relations. It has been shown that this can be done by fitting properties to polynomial functions. For many application programs, these polynomials are

$$\frac{C_p^\circ}{R} = a_1 + a_2 T + a_3 T^2 + a_4 T^3 + a_5 T^4 \tag{6.34}$$

$$\frac{H^\circ}{RT} = a_1 + \frac{a_2}{2} T + \frac{a_3}{3} T^2 + \frac{a_4}{4} T^3 + \frac{a_5}{5} T^4 + \frac{a_6}{6} T^5 \tag{6.35}$$

$$\frac{S^\circ}{RT} = a_1 \ln T + a_2 T + \frac{a_3}{2} T^2 + \frac{a_4}{3} T^3 + \frac{a_5}{4} T^4 + a_7 \tag{6.36}$$

In programs such as the NASA equilibrium code [33] or CHEMKIN [34], only the coefficients are stored. In addition, it has been found that to get acceptable fits, the full temperature range of interest must be split in two. Thus, 14 coefficients must be stored, as shown here for the molecule CH_4:

```
CH4                L 8/88C   1H   4   00   00G   200.000  3500.000   1000.000 1
 7.48514950E-02 1.33909467E-02-5.73285809E-06 1.22292535E-09-1.01815230E-13 2
-9.46834459E+03 1.84373180E+01 5.14987613E+00-1.36709788E-02 4.91800599E-05 3
-4.84743026E-08 1.66693956E-11-1.02466476E+04-4.64130376E+00 1.00161980E+04 4
```

6.2.3 Example 6.2

Consider one mole of water vapor, H_2O. If it is heated to a high enough temperature, then it can dissociate into O_2 and H_2. Here we solve the equilibrium problem at 3000 K numerically using Matlab.

Listing 6.1: Simple equilibrium script

```
1  close all
2  clear
3  clc
4  R = 0.0083144598; % kJ/mol/K
5  T = 3000; % K
6  % Species names
7  species = {'H2O' 'O2' 'H2'};
8  % Delta_fG^0 for included species at given T
9  % from JANNAF Tables
10 DfG0H2O = -77.163;
11 DfG0O2 = 0;
12 DfG0H2 = 0;
13 g0 = [ DfG0H2O DfG0O2 DfG0H2]; % kJ/mol
14 % Number of k type atoms in j molecule
15 nkj = [1    2    0       % oxygen
16        2    0    2]      % hydrogen
17 % Initial conditions - number of atoms
18 bk = [1     % oxygen atoms
19       2]    % hydrogen atoms
20 % Lower bounds on mole numbers
21 lb = [0 0 0]; % no mole numbers less than zero allowed
22 % Initial guesses for moles of each molecule
23 nj0 = [1e-3 1e-3 1e-3];
24 % Run minimization
25 gfun = @(nj)sum(nj.*(g0/(R*T) + log(nj/sum(nj))));
26 options = optimset('Algorithm','sqp');
27 [nj fval] = fmincon(gfun,nj0,[],[],nkj,bk,lb,[],[],options);
28 % Mole fractions
29 molfrac = (nj/sum(nj))';
30 fprintf('# Species Nj xj\n')
31 for i=1:numel(nj)
32 fprintf('%5d%10s%10.3g%10.3g\n',i,species{i},nj(i),molfrac(i))
33 end
```

Running the above script resulted in the mole fractions at 3000 K as shown in Table 6.1.

Table 6.1 Mole fractions predicted by Matlab script

H_2O	O_2	H_2
0.794	0.0687	0.137

6.2.4   The Equilibrium Constant

The discussion above provides a methodology for calculating the equilibrium composition of a reacting gas mixture. However, it is best suited to numerical calculations. For the case of one or two reactions, the equilibrium constant approach may be more useful.

We start by recalling that the equilibrium condition requires minimizing the Gibbs free energy

$$dG_{mix} = 0 \tag{6.37}$$

which, for a mixture of ideal gases, is

$$dG_{mix} = 0 = \sum_j^r dN_j g_j(T,p) = \sum_j^r N_j \ln[g_j(T,p) + RT\ln(p_j/p)] \tag{6.38}$$

Consider the reaction

$$aA + bB + \cdots \rightleftarrows eE + fF + \cdots \tag{6.39}$$

The change in the number of moles of each specie is directly proportional to its stoichiometric coefficient (a, b, etc.) times a progress of reaction variable α.

Note that we can write

$$\begin{aligned} dN_A &= -\alpha a \\ dN_B &= -\alpha b \\ &\vdots \\ dN_E &= +\alpha e \\ dN_F &= +\alpha f \end{aligned} \tag{6.40}$$

If we substitute this into the equation for dG_{mix}, we obtain

$$\Delta G_T = -RT \ln K_p \tag{6.41}$$

where

$$\Delta G_T = (eg_E + fg_F + \cdots - ag_A - bg_B - \cdots) \tag{6.42}$$

and

$$K_p = \frac{(p_E/p)^e (p_F/p)^f \ldots}{(p_A/p)^a (p_B/p)^b \ldots} \tag{6.43}$$

Here, K_p is the equilibrium constant.

It is common to define the Gibbs free energy of formation, $\Delta_f G^\circ$, as the value of ΔG at a reference pressure, typically the standard state pressure of 1 bar = 0.1 MPa. This makes it a function of temperature only. Then one can still use Eq. (5.18) as long as the equilibrium constant is written

$$K_p = \frac{(p_E/p_{ref})^e (p_F/p_{ref})^f \ldots}{(p_A/p_{ref})^a (p_B/p_{ref})^b \ldots} \tag{6.44}$$

where p_{ref} is the standard state pressure. Noting that

$$p_j = x_j p = \frac{N_j}{N_{tot}} p \tag{6.45}$$

we can write

$$K_p = \frac{N_E^e N_F^f \dots}{N_A^a N_B^b \dots} \left(\frac{P}{N_{tot}} \right)^{e+f+\dots-a-b-\dots} \tag{6.46}$$

Both $\Delta_f G^\circ$ and $\log_{10} K_p$ are tabulated in the JANNAF tables. They are related by

$$\Delta_f G^\circ(T) = -RT \ln K_p(T) \tag{6.47}$$

using the ideal gas relation. The molar concentration can be written (where the bracketed notation is common when discussing chemical reactions)

$$[C_j] = x_j \frac{p}{RT} = \frac{p_j}{RT} \tag{6.48}$$

so we can define an equilibrium constant based on molar concentrations:

$$K_p = K_c \left(\frac{RT}{p} \right)^{e+f+\dots-a-b-\dots} \tag{6.49}$$

where

$$K_c = \frac{[E]^e [F]^f \dots}{[A]^a [B]^b \dots} \tag{6.50}$$

6.2.5 Example 6.3

Repeat Example 6.2 using the equilibrium constant from the JANNAF tables.

Note that K_p is given for the formation of the compound from its elements in their natural state. Thus, the reaction we are concerned with is

$$H_2 + \frac{1}{2} O_2 \rightleftarrows H_2O$$

Write the reaction process as

$$H_2O \rightarrow xH_2O + yH_2 + zO_2$$

Carrying out an atom balance:

H	$2 = 2x + 2y$	$y = 1 - x$
O	$1 = x + 2y$	$z = (1 - x)/2$
N_{tot}	$x + y + z$	$1.5 - 0.5x$

Thus, using Eq. (6.44) and noting that

$$K_p = \frac{x}{(1 - x)[(1 - x)/2]^{1/2}} \left(\frac{p}{N_{tot}} \right)^{1-1/2-1}$$

from the JANNAF tables, $\log_{10} K_p = -1.344$ at 3000 K. Thus, $K_p = 0.0453$. Solving the equation for x, one obtains the same values as in Example 6.2.

6.2.6 The Principle of Detailed Balance

Now consider the reaction

$$A + B \underset{k_b}{\overset{k_f}{\rightleftarrows}} C + D \tag{6.51}$$

where k_f and k_b are called reaction rate coefficients. Suppose we wish to explore the time behavior of this reaction when starting from a non-equilibrium state, say all A and B but no C and D. It can be shown that the rate of destruction of A can be written as

$$\frac{d[A]}{dt} = -k_f[A][B] + k_b[C][D] \tag{6.52}$$

The reaction rate coefficients contain information about the nature of reacting collisions between molecules and are typically only a function of temperature. The concentration terms reflect the dependence on the frequency of collisions on concentration and thus pressure. If the mixture is in equilibrium, then the concentrations should be unchanging. Thus,

$$0 = -k_f[A][B] + k_b[C][D] \tag{6.53}$$

or

$$\frac{[C][D]}{[A][B]} = \frac{k_f(T)}{k_b(T)} = K_c(T) \tag{6.54}$$

This leads to the incredibly important principle of detailed balance. If the k's are a function of molecular structure and the translational energy is in equilibrium, then this relation should hold even when the mixture is not in chemical equilibrium. Using Eq. (6.65), only one of the two reaction rate coefficients is independently required, assuming the equilibrium constant is available.

6.3 Summary

We explored ideal gas mixtures in this chapter.

6.3.1 Non-reacting Mixtures

We started with non-reacting mixtures. For these mixtures, calculating the mixture properties is straightforward given the mole fractions of the components.

For the Gibbs representation in which temperature, pressure and mass, mole number, or molecular number are the independent variables, the following simple summation relations over all r species thus hold:

$$U = \sum_j^r U_j = \sum_j^r u_j N_j \tag{6.55}$$

$$H = \sum_{j}^{r} H_j = \sum_{j}^{r} h_j N_j \tag{6.56}$$

$$S = \sum_{j}^{r} S_j = \sum_{j}^{r} s_j N_j \tag{6.57}$$

for the extensive properties. Similarly for the normalized or specific properties:

$$u = \sum_{j}^{r} x_j u_j \tag{6.58}$$

$$h = \sum_{j}^{r} x_j h_j \tag{6.59}$$

$$s = \sum_{j}^{r} x_j s_j \tag{6.60}$$

$$c_v = \sum_{j}^{r} x_j c_{vj} \tag{6.61}$$

In the Gibbs representation, however, entropy is a function of both temperature and pressure. The s_j must be evaluated at the mixture temperature and the partial pressure,

$$s(T,p) = \sum_{j}^{r} x_j s_j(T,p_j) \tag{6.62}$$

$$s_2 - s_1 = c_p \ln \frac{T_2}{T_1} - R \ln \frac{p_2}{p_1} \tag{6.63}$$

6.3.2 Reacting Mixtures

For reacting mixtures we first discussed the general case of the equilibrium of a reacting mixture. Minimizing the Gibbs free energy, we derived the minimization function

$$G = \sum_{j}^{r} \left[g_j(T,p) + RT \ln x_j \right] \tag{6.64}$$

The minimization can easily be carried out numerically, as shown in Listing 6.1.

We then derived the equilibrium constant and showed how it can be used in Example 6.3. This was followed by a discussion of detailed balance:

$$\frac{[C][D]}{[A][B]} = \frac{k_f(T)}{k_b(T)} = K_c(T) \tag{6.65}$$

6.4 Problems

Reacting mixtures. For the following problems, "theoretical air" (or oxygen) means the percentage of air (or oxygen) that is provided compared to the stoichiometric amount. (You will need either the NIST database or your undergraduate thermodynamics books to do these problems, along with the NASA program CEA2.)

6.1 Liquid ethanol (C_2H_5OH) is burned with 150% theoretical oxygen in a steady-state, steady-flow process. The reactants enter the combustion chamber at 25°C and the products leave at 65°C. The process takes place at 1 bar. Assuming complete combustion (i.e. CO_2 and H_2O as products along with any excess O_2), calculate the heat transfer per kmole of fuel burned.

6.2 Gaseous propane at 25°C is mixed with air at 400 K and burned. 300% theoretical air is used. What is the adiabatic flame temperature? Again, assume complete combustion.

6.3 Repeat Problems 6.1 and 6.2 using CEA2. Compare the product composition using CEA2 with the assumption of complete combustion. What do you observe?

6.4 Using CEA2, calculate the equilibrium composition of a mixture of 1 mol of CO_2, 2 mol of H_2O, and 7.52 mol of N_2 at 1 bar and temperatures ranging from 500 to 3000 K in steps of 100 K. Then plot the mole fraction of NO as a function of temperature.

7 The Photon and Electron Gases

It is an interesting fact that both equilibrium radiation fields and electrons in metals can be treated as non-interacting ideal gases. Here we explore the consequences.

7.1 The Photon Gas

We have discussed the idea that electromagnetic radiation can, under certain circumstances, be thought of as being composed of photons, or quanta, that display particle-like characteristics. It would be useful if we could treat an electromagnetic field in an enclosure as a collection of monatomic particles and predict its equilibrium behavior using the principles of statistical mechanics. This is because many bodies emit radiation with a spectral distribution similar to that of an equilibrium field. A body that does so is said to be a "black body" and the radiation from the surface is completely characterized by the temperature. Indeed, in carrying out radiation heat transfer calculations, one usually expresses the properties of surfaces in terms of how closely they resemble a black body.

We can treat a collection of photons contained in an enclosure as an ideal monatomic gas except for two restrictions:

1. Photons are bosons and Bose–Einstein statistics must be used. However, photons do not interact with each other, so no approximation is made by neglecting interparticle forces.
2. Photons can be absorbed and emitted by the walls of the container, so that no constraint can be placed on the number of photons even in the grand canonical representation.

Recall that for a dilute assembly of bosons, the distribution of particles over the allowed energy states is

$$N_k = \frac{g_k}{e^{\gamma + \varepsilon_k/kT} - 1} \tag{7.1}$$

However, this expression was derived including the constraint on N that is now removed and $\gamma = -\mu/kT = 0$. Thus,

$$N_k = \frac{g_k}{e^{\varepsilon_k/kT} - 1} \tag{7.2}$$

If we assume that the differences between energy levels are so small that energy varies continuously, we can write

$$dN = \frac{dg}{e^{\varepsilon/kT} - 1} \tag{7.3}$$

where dg is the degeneracy for the energy increment ϵ to $\epsilon + d\epsilon$. As we derived in our discussion of translational energy, the degeneracy is

$$dg = 2\left(\frac{4\pi n^2 dn}{8}\right) \tag{7.4}$$

where the factor of 2 arises because an electromagnetic field can be polarized in two independent directions.

For a molecular gas, we related dn to $d\epsilon$ using Newtonian physics, namely that

$$\varepsilon = \frac{mc^2}{2} \text{ or } p^2 = 2m\varepsilon \tag{7.5}$$

However, photons are relativistic, and de Broglie's relation (where here c is the speed of light)

$$p = \frac{\varepsilon}{c} \tag{7.6}$$

must be used. Therefore,

$$\varepsilon^2 = \frac{h^2c^2}{4V^{2/3}}n^2 \tag{7.7}$$

and

$$dg = \frac{8\pi V}{h^3c^3}\varepsilon^2 d\varepsilon \tag{7.8}$$

Substituting this into the expression for dN and noting that $\varepsilon = h\nu$,

$$\frac{dN}{V} = \frac{8\pi}{c^3}\frac{\nu^2}{(e^{h\nu/kT} - 1)}d\nu \tag{7.9}$$

This is the number of photons (per unit volume) in the frequency range ν to $\nu + d\nu$. We seek the spectral energy density u_ν:

$$u_\nu d\nu = h\nu\frac{dN}{V} \tag{7.10}$$

Thus,

$$u_\nu = \frac{8\pi h\nu^3}{c^3}\frac{1}{(e^{h\nu/kT} - 1)} \tag{7.11}$$

This is Planck's Law.

The limits for low and high frequency are readily obtained. For large ν, $h\nu/kT >> 1$ and we get the Wien formula:

$$u_\nu \cong \frac{8\pi h\nu^3}{c^3}e^{-h\nu/kT} \tag{7.12}$$

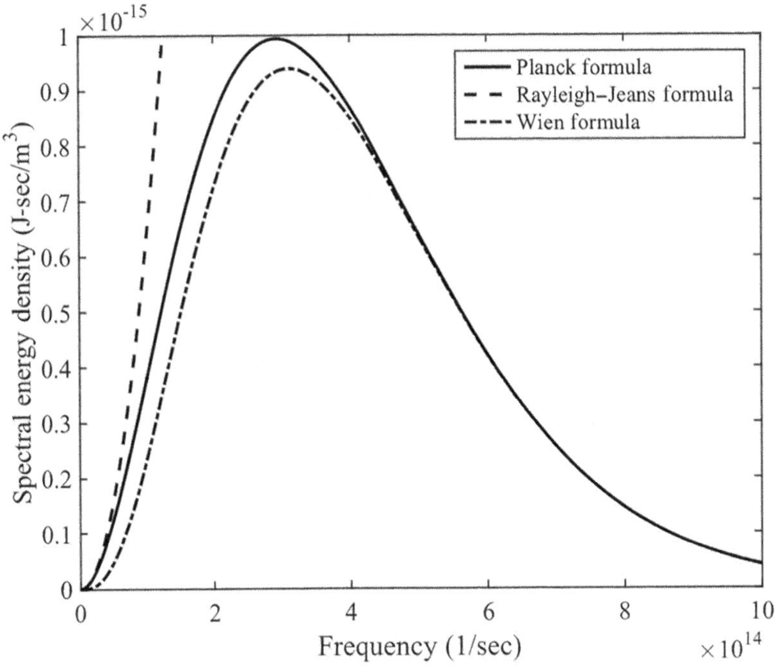

Figure 7.1 Spectral distribution of blackbody radiation.

For small ν, $h\nu/kT << 1$ and we get the Rayleigh–Jeans formula:

$$u_\nu \cong \frac{8\pi\nu^2 kT}{c^3} \tag{7.13}$$

These functions are plotted in Fig. 7.1.

The wavelength of maximum energy or intensity is known as Wien's displacement law. It can be obtained by taking the derivative of Planck's Law and setting it to zero. The result is

$$\lambda = \frac{b}{T} \tag{7.14}$$

where $b = 2.898 \times 10^{-3}$ m K.

The total energy per unit volume is

$$u = \int_0^\infty u_\nu d\nu = \frac{8\pi h}{c^3}\int_0^\infty \frac{\nu^3}{(e^{-h\nu/kT} - 1)} d\nu = \frac{8\pi^5}{15}\frac{(kT)^4}{(hc)^3} \tag{7.15}$$

We could have obtained this relation by first evaluating the partition function, thereby obtaining the fundamental relation. Energy is then one of the equations of state in the grand canonical representation. For a boson and with $\gamma = 0$:

$$\ln Q = \sum_k \ln \frac{1}{(1 - e^{-\beta\varepsilon_k})} \cong \int_0^\infty g(\nu)\ln\frac{1}{(1 - e^{-h\nu/kT})} d\nu = \frac{8\pi^5 V}{45(hc/kT)^3} \tag{7.16}$$

Therefore, the fundamental relation becomes

$$S[1/T, 0] = \frac{8k\pi^5 V}{45(hc/kT)^3} = \frac{pV}{T} \tag{7.17}$$

The energy we have already derived. The pressure is

$$p = \left.\frac{\partial S[1/T]}{\partial V}\right)_V = \frac{8\pi^5}{45(hc)^3}(kT)^4 \tag{7.18}$$

To obtain the number of photons requires differentiating with respect to μ before setting $d\mu = 0$ (see Knuth [5], p. 101):

$$N = -\left.\frac{\partial S[1/T, -\mu/T]}{\partial \mu/T}\right)_{1/T,V} = \frac{\pi^2}{6}\frac{16\pi V}{(hc/kT)^3} \tag{7.19}$$

The entropy for the photon gas is

$$S = \frac{32k\pi^5 V}{45(hc/kT)^3} \tag{7.20}$$

Finally, we can also derive the important Stefan–Boltzmann relation used in heat transfer. For an ideal gas, the flux of particles across any given plane is $n\bar{c}/4$. This relation was derived by integration of the Maxwellian velocity distribution function. By analogy, we can show that the energy flux of photons (energy per unit time and area) is

$$q = \frac{2\pi^5 k^4}{15h^3c^2}T^4 = \sigma T^4 \tag{7.21}$$

Therefore,

$$\sigma = \frac{2\pi^5 k^4}{15h^3c^2} = 5.670373(21) \times 10^{-8}\ \mathrm{W\ m^{-2}\ K^{-4}} \tag{7.22}$$

is the Stefan–Boltzmann constant.

7.1.1 Example 7.1

Calculate the energy density contained between 2 and 5 µm by a black body at 200°C.

This involves integrating the blackbody function over the indicated range. The limits of integration are given in units of wavelength, 2 and 5 µm. To convert to frequency, we use the relation

$$\nu = \frac{c}{\lambda}$$

where c is the speed of light. Thus, $\nu_1 = 6 \times 10^{15}\ \mathrm{sec}^{-1}$ and $\nu_1 = 1.5 \times 10^{16}\ \mathrm{sec}^{-1}$.

This can easily be integrated in Matlab or Mathematica:

$$u = \int_{\nu_1}^{\nu_2} u_\nu d\nu = \frac{8\pi h}{c^3}\int_{\nu_1}^{\nu_2} \frac{\nu^3}{(e^{-h\nu/kT} - 1)} d\nu = 3.792 \times 10^{-5}\ \mathrm{J/m^3} \tag{7.23}$$

This represents 13.3% of the total energy.

7.2 The Electron Gas

Metals are excellent conductors of heat and electricity. It is observed that electrical current in a metal involves the flow of electrons, but not ions. A simple model proposed by Paul Drude in 1900 assumed that when the atoms of a metallic element are brought together to form a metal, the valence electrons become detached and are free to move within the metal lattice. If we assume that the electrons do not interact with one another, then they collectively must act like a monatomic ideal gas.

If free electrons do act like an ideal gas, each one would contribute $3/2kT$ to the total energy, so if N_0 is the number of free electrons:

$$U = N_0 \frac{3}{2} kT \tag{7.24}$$

The specific heat per electron would then be

$$c_v = \frac{3}{2} k \tag{7.25}$$

The number density of free electrons depends on the specific metallic atom. Table 7.1 shows experimentally determined values for a number of metals. However, for many metals, there is about one free valence electron per atom.

The Drude model is not very accurate and a fully quantum analysis is more appropriate. Electrons must satisfy the Pauli exclusion principle. Therefore, they will fill the metal lattice in increasing energy order. This is illustrated in Fig. 7.2. Even at absolute zero, not all the electrons will have zero energy. The maximum electron energy at 0 K is called μ_0, the Fermi energy.

The quantum state number distribution for a fermion is

$$N_k = \frac{1}{e^{(\varepsilon_k - \mu)/kT} + 1} \tag{7.26}$$

Table 7.1 Electronic properties of metals

Metal	Free electron number density $10^{22}/\text{cm}^3$	Fermi level (eV)	Specific heat (kJ/kg K)
Li	4.6	4.74	3.57
Na	2.5	3.24	1.21
K	1.3	2.12	0.75
Cu	8.47	7.00	0.39
Ag	5.86	5.49	0.13
Au	5.9	5.53	0.23
Fe	17.0	11.1	0.45
W	—	4.5	0.13
Al	11.7	18.1	0.91
Pt	6.6	5.30	0.13
Zn	13.2	9.47	0.39

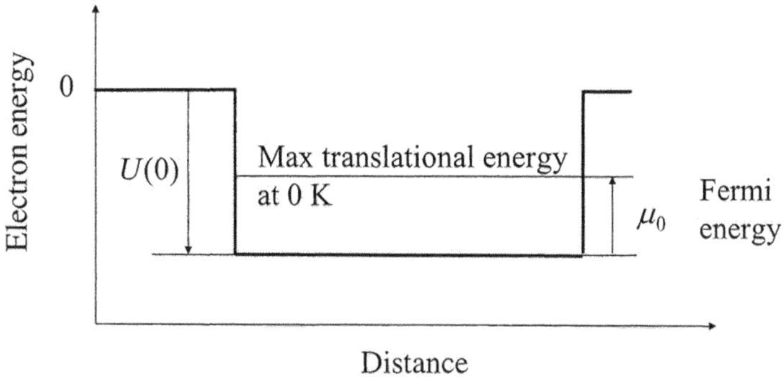

Figure 7.2 Electron energy distribution in a metal matrix.

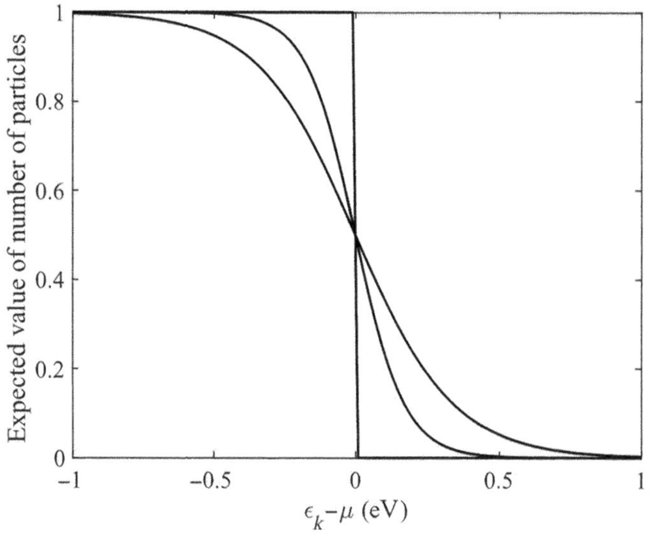

Figure 7.3 Electron energy distribution.

and in the continuum limit the single-electron energy distribution is (see Fig. 7.3)

$$f(\varepsilon) = \frac{1}{e^{(\varepsilon-\mu)/kT} + 1} \tag{7.27}$$

The number of states with energy between ϵ and $\epsilon + d\epsilon$ is the degeneracy

$$dg = 4\pi V \left(\frac{2m_e}{h^2}\right)^{3/2} \sqrt{\varepsilon} d\varepsilon \tag{7.28}$$

The factor of 4 arises because electrons can each have two spin states. Thus, the equivalent expression for particles in a box [Eq. (5.19)] must be multiplied by 2. Then,

$$N(\varepsilon)d\varepsilon = \left[4\pi V \left(\frac{2m_e}{h^2}\right)^{3/2}\right] \frac{\sqrt{\varepsilon}}{e^{(\varepsilon-\mu)/kT} + 1} d\varepsilon \tag{7.29}$$

Integrating the energy distribution at 0 K, the number of free electrons equals

$$N = 4\pi \frac{8\pi}{3}\left(\frac{2m_e}{h^2}\right)^{3/2} V \int_0^{\mu_0} \varepsilon^{1/2} d\varepsilon = 4\pi \frac{8\pi}{3}\left(\frac{2m_e}{h^2}\right)^{3/2} V(\mu_0)^{3/2} \tag{7.30}$$

Solving for the Fermi energy, we obtain

$$\mu_0 = \frac{h^2}{2m_e}\left(\frac{3}{8\pi}\right)^{2/3}\left(\frac{N}{V}\right)^{2/3} \tag{7.31}$$

Values of the Fermi energy are typically in the range of 1–5 eV. We can define a Fermi temperature as

$$T_F = \frac{\mu_0}{k} \tag{7.32}$$

and since 1 eV = 11,600 K, the Fermi temperatures are typically very large. As a consequence, electron gases behave as though they are at a very low temperature.

The total electron energy at absolute zero is

$$U_0 = \int_0^{\infty} \epsilon dN(\epsilon) = \frac{\pi}{2}\left(\frac{8m_e V^{2/3}}{h^2}\right)^{3/2} \int_0^{\epsilon_0} \epsilon^{3/2} d\epsilon = \frac{3}{5} N\mu_0 \tag{7.33}$$

We can easily determine the fundamental relation at 0 K. It is simply the ideal gas result in the Helmholtz representation

$$F_0 = NU(0) + \frac{3}{40} N \frac{h^2}{m_e}\left(\frac{3N}{\pi V}\right)^{2/3} \tag{7.34}$$

At higher temperature things would be simplified if the Maxwell–Boltzmann limit applied. However, for copper at its normal melting point, for example,

$$\left(\frac{2\pi m_e kT}{h^2}\right)^{3/2} \frac{V}{N} \approx 1.42 \times 10^{-3} << 1 \tag{7.35}$$

Therefore, the full Fermi–Dirac treatment must be followed and a series expansion approach taken:

$$F = F_0 \left[1 - \frac{5\pi^2}{12}\left(\frac{T}{T_F}\right)^2 + \cdots\right] \tag{7.36}$$

This leads to

$$U = U_0 \left[1 + \frac{5\pi^2}{12}\left(\frac{T}{T_F}\right)^2 - \cdots\right] \tag{7.37}$$

where F_0 and U_0 are reference values of F and U, respectively.

Since in all practical cases $T \ll T_F$, the specific heat per electron becomes

$$c_v \cong \frac{\pi^2}{2} k \frac{T}{T_F} \tag{7.38}$$

This is far less than the ideal gas value of $\frac{3}{2}kT$.

7.2.1 Example 7.2

Calculate the electron number density for tungsten.

The number density of electrons is related to the Fermi energy by the relation

$$n = 4\pi \frac{8\pi}{3} \left(\frac{2m_e}{h^2} \right)^{3/2} (\mu_0)^{3/2}$$

The Fermi energy for tungsten is 4.5 eV, so that

$$n = 5.45 \times 10^{29}\ \mathrm{m}^{-3}$$

7.2.2 Example 7.3

Plot the electron contribution to the specific heat of tungsten as a function of temperature from 200 to 1000 K.

$$\frac{c_v}{k} \cong \frac{\pi^2}{2} \frac{T}{T_F}$$

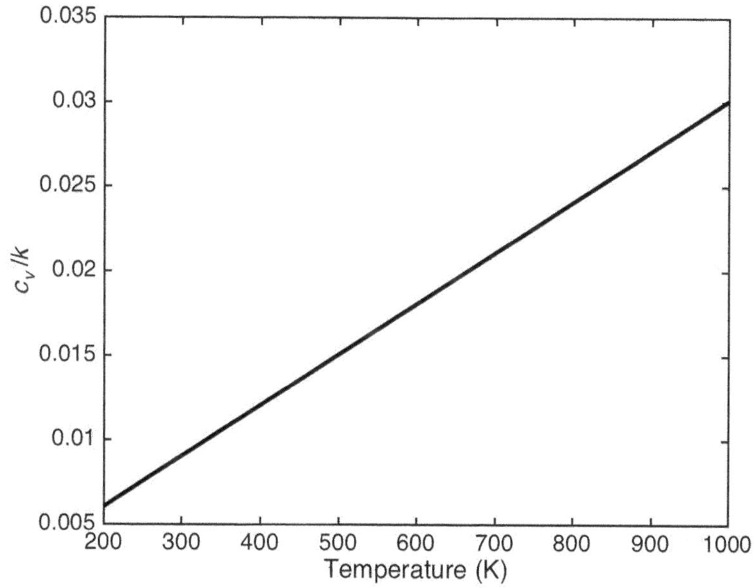

Figure 7.4 Specific heat as a function of temperature.

7.3 Summary

In this chapter we discussed the photon and electron gases. What is unique about both is that they can be treated as ideal gases subject to some modifications.

7.3.1 Photon Gas

Two special conditions apply to photons:

1. Photons are bosons and Bose–Einstein statistics must be used. However, photons do not interact with each other so that no approximation is made by neglecting inter-particle forces.
2. Photons can be absorbed and emitted by the walls of the container, so that no constraint can be placed on the number of photons even in the grand canonical representation.

The important results are Planck's blackbody function

$$u_\nu = \frac{8\pi h\nu^3}{c^3}\frac{1}{(e^{h\nu/kT} - 1)} \tag{7.39}$$

and the Stefan–Boltzmann constant

$$\sigma = \frac{2\pi^5 k^4}{15h^3c^2} = 5.670373(21) \times 10^{-8}\ \mathrm{W\,m^{-2}\,K^{-4}} \tag{7.40}$$

7.3.2 Electron Gas

Electrons are fermions and Fermi–Dirac statistics must be followed. This results in the electron number distribution

$$N(\varepsilon)d\varepsilon = \left[4\pi V\left(\frac{2m_e}{h^2}\right)^{3/2}\right]\frac{\sqrt{\varepsilon}}{\mathrm{e}^{(\varepsilon-\mu)/kT} + 1}d\varepsilon \tag{7.41}$$

This is illustrated in Fig. 7.3. Note that because electrons satisfy the Pauli exclusion principle, at a given energy the maximum number of electrons allowed is unity.

We also derived the electron contribution to the specific heat:

$$c_v \cong \frac{\pi^2}{2}k\frac{T}{T_F} \tag{7.42}$$

The magnitude of this contribution is far less than the contribution from vibrational motion within a sold (see Chapter 10).

7.4 Problems

7.1 Calculate and plot the photon pressure due to blackbody radiation [Eq. (7.18)] as a function of temperature from 200 to 2000 K.

7.2 The intensity distribution of solar radiation peaks at a wavelength of approximately 500 nm. Assuming the Sun radiates like a black body, calculate its surface temperature. Calculate the flux of photons leaving the Sun.

7.3 Plot the number distribution of electrons versus ϵ [Eq. (5.134)] for a 1-cm^3 sample of copper at 0 K and 300 K. Use a density of 8.96 gm/cm^3 and assume one conduction electron per atom.

7.4 Calculate the free electron number density for iron. Compare the value to that in Table 7.1.

7.5 Plot the specific heat of iron as a function of temperature from 200 to 1000 K.

8 Dense Gases

So far we have explored the situation where the density is low enough that we can treat the particles (atoms/molecules) as non-interacting. Once we relax that approximation, things get a bit more complicated and interatomic/molecular forces must be taken into account. Recall the character of the interatomic potential function as illustrated in Fig. 2.1. It is typically the case that as two atoms (or molecules) are brought closer together, they first experience an attractive force. This is called the van der Waals force. However, if the positively charged nuclei are brought too closely together, a very strong force of repulsion develops. The presence of these forces will have a significant effect on the system energy, which can be expressed as

$$U_j = \sum_{i=1}^{N} (\epsilon_i)_{int} + \sum_{i=1}^{N} \frac{|p_i|^2}{2m} + \phi(\vec{r}_1, \vec{r}_2, \ldots) \tag{8.1}$$

where the first term is the internal energy, the second is the translational energy and the third is the energy associated with intermolecular forces. The (canonical) partition function then becomes

$$Q = \sum_j e^{-\beta U_j} = \sum_j e^{-\beta U_j^{int}} \cdot \sum_j e^{-\beta(U_j^{trans} + \phi_j)} \tag{8.2}$$

where we have factored out the internal energy term. Since

$$U_j^{int} = \sum_{i=1}^{N} \epsilon_i^{int} \tag{8.3}$$

we have

$$\sum_j e^{-\beta U_j^{int}} = \prod_{i=1}^{N} \sum_k e^{-\beta \epsilon_k^{int}} \tag{8.4}$$

Note that the sum is the internal energy partition function. Therefore,

$$\sum_j e^{-\beta U_j^{int}} = q_{int}^N \tag{8.5}$$

and

$$Q = q_{int}^N \cdot \sum_j e^{-\beta(U_j^{trans} + \phi_j)} \tag{8.6}$$

Because translation states are so closely spaced, one can replace the sum with an integral:

$$\sum_j e^{-\beta(U_j^{trans}+\phi_j)} = \frac{1}{N!\,h^{3N}} \int \cdots \int e^{-\beta H} d\vec{p}_1 d\vec{p}_2 \ldots d\vec{p}_N d\vec{r}_1 d\vec{r}_2 \ldots d\vec{r}_N \tag{8.7}$$

where H is the classical Hamiltonian

$$H = \frac{1}{2m} \sum_{n=1}^{N} (p_{xn}^2 + p_{yn}^2 + p_{zn}^2) + \phi(\vec{r}_1, \vec{r}_2 \ldots \vec{r}_N) \tag{8.8}$$

The integral over momentum is just the translational energy

$$\langle \varepsilon_{tr} \rangle \left(\frac{2\pi mkT}{h^2} \right)^{\frac{3}{2}N} \tag{8.9}$$

while the integral over space is

$$Z_\phi = \int \cdots \int e^{-\phi/kT} dr_1 dr_2 \ldots dr_N \tag{8.10}$$

We call this the "configuration integral" because it depends on the configuration (i.e. the location of all the atoms and molecules).

The canonical partition function then becomes

$$Q = q_{int}^N \frac{1}{N!} \left(\frac{2\pi mkT}{h^2} \right)^{\frac{3}{2}N} Z_\phi \tag{8.11}$$

Finally, evaluating the canonical Massieu function, we obtain the fundamental relation

$$\frac{S[1/T]}{k} = -\frac{F}{kT} = -\ln Q = N \left[\ln q_{int} - (\ln N - 1) + \frac{3}{2} \ln \left(\frac{2\pi mkT}{h^2} \right) \right] + \ln Z_\phi \tag{8.12}$$

8.1 Evaluating the Configuration Integral

Evaluating the configuration integral analytically requires some simplifying assumptions. We start with these two:

- The orientation of molecules is unimportant, so that the force between any two molecules is a function of distance only. (This is not true for highly non-symmetric molecules.)
- The potential energy is the sum of the pairwise potential energies between molecules. (This only strictly applies to a gas that is not too dense.)

The first assumption greatly simplifies the calculation of intermolecular forces. Essentially we treat molecules as spherically symmetric points and the intermolecular potentials depend only on the distance between the points. The second assumption simplifies calculating the force on a given molecule by all other molecules. Thus we can write

$$\phi(\vec{r_1}, \vec{r_2} \ldots) = \frac{1}{2}\sum_{i \neq j} \phi_{ij}(r_ij) = \sum_{1 \leq i < j \leq N} \phi_{ij}(r_ij) \tag{8.13}$$

where

$$r_{ij} \equiv |\vec{r_i} - \vec{r_j}| \tag{8.14}$$

We can simplify by introducing the function

$$f_{ij}(r_{ij}) = e^{-\phi_{ij}(r_{ij})} - 1 \tag{8.15}$$

in which case we can write

$$e^{-\phi/kT} = \prod_{1 \leq i < j \leq N} e^{-\phi_{ij}(r_{ij})/kT} = \prod_{1 \leq i < j \leq N} (1 + f_{ij}) \tag{8.16}$$

Next we assume that the gas is only weakly interacting. If that is the case, then f_{ij} must be much less than one. (Alternatively put, we will explore the result when $f_{ij} \ll 1$.) This allows us to drop the higher-order terms in the expansion of Eq. (8.16), in which case

$$e^{-\phi/kT} \simeq 1 + \sum_{1 \leq i < j \leq N} f_{ij} \tag{8.17}$$

and the configuration integral becomes

$$Z_\phi = \int \cdots \int \left(1 + \sum_{1 \leq i < j \leq N} f_{ij} \right) dr_1 dr_2 \ldots dr_N \tag{8.18}$$

Noting that

$$\int_{-\infty}^{\infty} dr = V \tag{8.19}$$

then

$$Z_\phi = V^N + \int \cdots \int \sum_{1 \leq i < j \leq N} f_{ij} dr_1 dr_2 \ldots dr_N \tag{8.20}$$

However, since f_{ij} depends only on r_{ij},

$$Z_\phi = V^N + V^{N+1} 2\pi N^2 \int_{-\infty}^{\infty} f(r) dr \tag{8.21}$$

where we have assumed that all the f_{ij} are the same function.

Finally, it is conventional to write Z_ϕ in the form

$$Z_\phi = V^N \left[1 - \frac{N^2 B(T)}{V} \right] \tag{8.22}$$

where $B(T)$ is called the second virial coefficient:

$$B(T) = -2\pi \int_{-\infty}^{\infty} f(r) dr \tag{8.23}$$

If we were to drop the weakly interacting assumption and start admitting higher-order terms in Eq. (8.17), then we could more generally write

$$Z_\phi = V^N \left[1 - \frac{N^2 B(T)}{V} + \text{higher-order terms} \right] \tag{8.24}$$

8.2 The Virial Equation of State

Once we know the configuration integral it turns out that we can derive an equation of state for pressure, called the virial equation of state. In the canonical representation:

$$dS[1/T] = -U d(1/T) + \frac{p}{T} dV - \frac{\mu}{T} dN \tag{8.25}$$

Recalling that

$$\frac{p}{T} = k \ln Q \tag{8.26}$$

and since

$$S[1/T] = k \ln Q \tag{8.27}$$

then

$$p = kT \left.\frac{\partial \ln Q}{\partial v}\right)_{1/T,N} \tag{8.28}$$

The only term in $\ln Q$ that contains V is $\ln Z_\phi$, so

$$p = kT \left.\frac{\partial \ln Z_\phi}{\partial V}\right)_{1/T,N} \tag{8.29}$$

Now, $\ln Z_\phi$ is in the form of $\ln(1+x)$ where x is small, and

$$\ln(1+x) = x - \frac{x^2}{2} + \cdots \tag{8.30}$$

Using this fact we can write p as

$$p = \frac{NkT}{V} + \frac{N^2 kTB(T)}{V^2} + \cdots \tag{8.31}$$

or

$$p = \frac{kT}{v} + \frac{kTB(T)}{v^2} + \cdots \tag{8.32}$$

where $v = V/N$.

In molar units, $R = N_A k$ where N_A is Avogadro's number. Thus,

$$p = \frac{RT}{v} + \frac{RTN_A B(T)}{v^2} + \cdots \tag{8.33}$$

where $v = VN_A/N$ in this case.

This equation is more commonly expressed in the following form:

$$p = \frac{RT}{v} + \frac{a(T)}{v^2} + \frac{b(T)}{v^3} + \cdots \tag{8.34}$$

with the coefficients determined empirically. Most other real gas equations of state are similarly expansions about the ideal gas limit. Examples are the van der Waals, Beattie–Bridgeman, Redlich–Kwong, and Benedict–Webb–Rubin equations.

8.3 Other Properties

Since we have an expression for the partition function in terms of the configuration integral, we can calculate the other thermodynamic properties. Recall that

$$dS[1/T] = -Ud(1/T) + \frac{p}{T}dV - \frac{\mu}{T}dN \tag{8.35}$$

Therefore,

$$U = -\left.\frac{\partial S[1/T]}{\partial[1/T]}\right)_{V,N} \tag{8.36}$$

and the energy becomes

$$U = \frac{3}{2}NkT + kT^2\left(\frac{\partial \ln Z_\phi}{\partial T}\right)_{NV} = \frac{3}{2}NkT + \bar{U} \tag{8.37}$$

where

$$\bar{U} = \frac{\int\cdots\int \phi e^{-\phi/kT} dr_1 dr_2 \ldots dr_N}{Z_\phi} \tag{8.38}$$

can be considered as the configuration energy.

8.4 Potential Energy Functions

To evaluate the configuration integral using Eq. (8.21), for example, requires knowing the function $\phi(r)$, which is known as the potential energy function. In general, the potential energy function is a somewhat complex function of distance between the two particles, even for collisions between atoms that are spherically symmetric. Coming to our assistance, however, is the fact that this function only appears in an integral, which is somewhat insensitive to the exact shape of the potential. As a result, there have been many algebraic forms proposed for use.

The simplest form is that for a rigid sphere (Fig. 8.1), where σ is the line-of-center distance at which two particles are touching. Also called the billiard ball model, it assumes no forces between the particles until they hit a solid surface. Also shown in Fig. 8.1 are the weakly attracting, square well, and Lennard–Jones potentials.

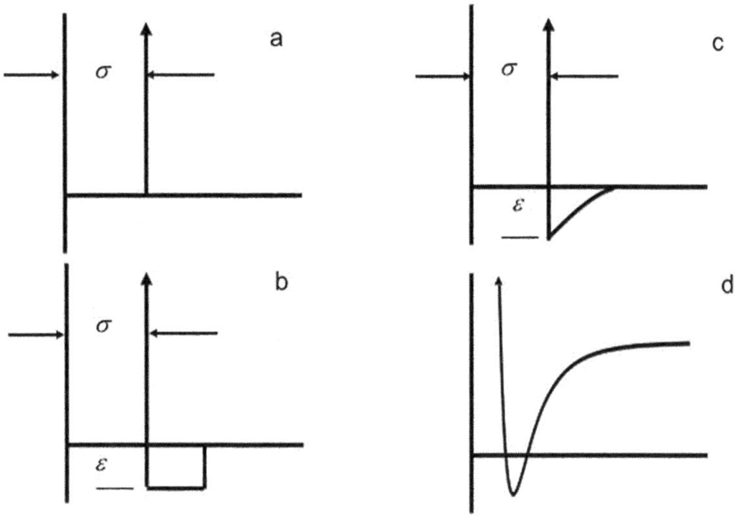

Figure 8.1 Rigid sphere, square well, Sutherland (weakly attractive sphere), and Lennard–Jones potential functions (clockwise around figure).

8.4.1 Example 8.1

Calculate the second virial coefficient for the rigid sphere potential and use the result in Eq. (8.33).

Recall that the parameter $B(T)$ is of the form

$$B(T) = -2\pi \int_0^\infty f(r)dr \tag{8.39}$$

For the rigid sphere model this can be written as

$$B(T) = -2\pi \left\{ \int_0^\sigma (0-1)r^2 dr + \int_\sigma^\infty (1-1)r^2 dr \right\} = 2\pi \frac{\sigma^3}{3} \tag{8.40}$$

Using this result in Eq. (8.33), we obtain

$$p = \frac{RT}{v} + \frac{RTB(T)}{v^2} \tag{8.41}$$

or [it is common to denote $B(T)$ as b]

$$\frac{pv}{RT} = \frac{v+b}{v} \tag{8.42}$$

If $b << v$, then

$$p = \frac{RT}{(v-b)} \tag{8.43}$$

This is the first term in the van der Waals equation. Note that the volume occupied by a spherical molecule is

$$\frac{4\pi}{3}r^3 = \frac{4\pi}{3}\left(\frac{\sigma}{2}\right)^3 \tag{8.44}$$

This is called the co-volume. It is the volume excluded to other molecules.

The rigid sphere model has one parameter, σ. While it obviously has the wrong shape, by comparing with experimental pvT data, a value for σ can be obtained by regression. In some cases, and for not too broad a range of conditions, the resulting equation of state can work reasonably well. However, the ability to fit data over a wider range of conditions requires a potential energy function that more closely approximates reality. Take the weakly attracting sphere or Sutherland potential, for example, which is a three-parameter function, σ, ε, and γ (γ is typically > 3):

$$\phi(r) = \begin{cases} \infty & r \leq \sigma \\ -\epsilon\left(\frac{\sigma}{r}\right)^\gamma & \sigma < r \end{cases}$$

8.4.2 Example 8.2

Calculate the second virial coefficient for the weakly attracting Sutherland potential and use the result in Eq. (8.33). (Note, weakly attracting means that $\exp(\frac{\epsilon}{kT}(\frac{\sigma}{r})^\gamma) << 1$.)

For this function, $B(T)$ becomes

$$B(T) = -2\pi\left\{\int_0^\sigma r^2 dr + \int_\sigma^\infty (1 - e^{\frac{cr^{-\gamma}}{kT}})r^2 dr\right\} \tag{8.45}$$

However, $\exp\left(\frac{cr^{-\gamma}}{kT}\right) \sim 1 + \frac{cr^{-\gamma}}{kT}$ so that

$$B(T) = 2\pi\left(\frac{\sigma^3}{3} - \frac{c}{kT}\frac{\sigma^{3-\gamma}}{3-\gamma}\right) \tag{8.46}$$

This equation can be written as

$$B(T) = b - \frac{a}{RT} \tag{8.47}$$

and if $b << a/RT$, then

$$p = \frac{RT}{(v-b)} - \frac{a}{v^2} \tag{8.48}$$

This, of course, is the van der Waals equation.

One can estimate the parameters a and b by setting $(\partial P/\partial v)_{T_c} = (\partial^2 P/\partial v^2)_{T_c} = 0$. This results in a solution

$$a = \frac{9RT_c v_v}{8} \tag{8.49}$$

$$b = \frac{v_c}{3} \tag{8.50}$$

Setting P, t, and v to their critical values in Eq. (8.48) results in

$$P_c = \frac{3RT_c}{8v_c} \tag{8.51}$$

This allows one to express a and b in terms of P_c and T_c, which generally results in a better fit to data:

$$a = \frac{27R^2T_c^2}{64P_c} \tag{8.52}$$

$$b = \frac{RT_c}{8P_c} \tag{8.53}$$

We know that while the van der Waals equation is very useful for illustrating the role of attractive and repulsive forces, it is not very accurate. An even better form of the potential energy function that mimics the shape of many real potential functions is the Lennard–Jones potential, also known as the 6–12 potential:

$$\phi(r) = 4\epsilon_{ij}\left[\left(\frac{\sigma_{ij}}{r}\right)^{12} - \left(\frac{\sigma_{ij}}{r}\right)^{6}\right] \tag{8.54}$$

Here, σ_{ij} is the interatomic distance at which the attractive and repulsive forces exactly balance and ϵ is the well depth where ϕ is a minimum. Lennard–Jones is commonly used, especially for calculating transport properties. However, it, and other more realistic potential functions, do not result in an algebraic form for $B(T)$, rather the integral expression

$$B(T) = 2\pi \int_0^\infty \left[1 - \exp\left\{-\frac{4\epsilon}{kT}\left[\left(\frac{\sigma}{r}\right)^{12} - \left(\frac{\sigma}{r}\right)^{6}\right]\right\}\right] r^2 dr \tag{8.55}$$

must be evaluated numerically.

It is usual to normalize the parameters as

$$T^* \equiv \frac{kT}{\epsilon} \tag{8.56}$$

and

$$B^* \equiv \frac{B}{b} \tag{8.57}$$

where $b = \frac{2\pi}{3}\sigma^3$, then

$$B^* = B^*(T^*) \tag{8.58}$$

This function can be tabulated for practical use. See, for example, Appendix E.

8.5 Other Equations of State

We have explored the underlying physics of non-ideal gas behavior using relatively simple approximations. These include pairwise potential functions and simple analytic forms, and have allowed derivation of van der Waals equation. Unfortunately, van der Waals equation, while very useful for illustrating the important physics of pvT behavior, is not very accurate for real substances. Furthermore, even today when very powerful quantum mechanics software is available, the difficulties imposed by the three-dimensional structures of most interesting substances means that empirically based methods have historically been used to provide more accurate equations of state, including ones that extend into liquid and supercritical regions. Here we present three well-known such equations.

The Redlich–Kwong equation was published in 1949 [35]. It is a two-parameter equation of state, and does well at small to moderate densities and when $T > T_c$:

$$P = \frac{RT}{v-b} - \frac{a}{T^{1/2}v(v+b)} \tag{8.59}$$

Values of the parameters a and b can be found in many thermodynamics textbooks. In the absence of experimental data, one can estimate the parameters using critical data. As illustrated for van der Waals equation, setting $(\partial P/\partial v)_{T_c} = (\partial^2 P/\partial v^2)_{T_c} = 0$ and using Eq. (8.80) one obtains

$$a = 0.42748\frac{R^2 T_c^{2.5}}{P_c} \tag{8.60}$$

$$b = 0.08664\frac{RT_c}{P_c} \tag{8.61}$$

Another two-parameter equation of state is Peng–Robinson [36], developed for vapor–liquid equilibrium calculations (see [36] for definitions):

$$P = \frac{RT}{v-b} - \frac{a\alpha}{v(v+b)+b(v-b)} \tag{8.62}$$

Again, following the example above, one can estimate the parameters as a function of the critical parameters:

$$a = 0.45724\frac{R^2 T_c^2}{P_c} \tag{8.63}$$

$$b = 0.07780\frac{RT_c}{P_c} \tag{8.64}$$

The Benedict–Webb–Rubin equation of state [37, 38] is a more complex virial expansion-type equation. It contain eight constants (see [37, 38] for definitions):

$$P = \frac{RT}{v} + \frac{B_0 RT - A_0 - C_0/T^2}{v^2} + \frac{bRT - a}{v^3} + \frac{a\alpha}{v^6} + \frac{c}{v^3 T^2}\left(1 + \frac{\gamma}{v^2}\right)\exp\left(\frac{-\gamma}{v^2}\right) \tag{8.65}$$

Parameters for a number of species are given in Cooper and Goldfrank [39], Perry and Green [8], and Hirschfelder, Curtiss, and Bird [14].

In practice, engineers are most likely to turn to digital compilations to obtain property data for real substances.

8.6 Summary

In this chapter we considered thermodynamic behavior when intermolecular forces become important. We introduced the concept of the configuration integral, and explored the consequences:

$$Z_\phi = \int \cdots \int e^{-\phi/kT} dr_1 dr_2 \ldots dr_N \tag{8.66}$$

The canonical partition function becomes

$$Q = q_{int}^N \frac{1}{N!} \left(\frac{2\pi mkT}{h^2} \right)^{\frac{3}{2}N} Z_\phi \tag{8.67}$$

8.6.1 Evaluating the Configuration Integral

Evaluating the configuration integral requires knowledge of the intermolecular potential function. If we make some simplifying assumptions:

- The orientation of molecules is unimportant, so that the force between any two molecules is a function of distance only. (This is not true for highly non-symmetric molecules.)
- The potential energy is the sum of the pairwise potential energies between molecules. (This only strictly applies to a gas that is not too dense.)
- The interactions are weak (i.e. small departures from ideal gas behavior).

then we can expand the exponential term as

$$e^{-\phi/kT} \simeq 1 + \sum_{1 \leq i < j \leq N} f_{ij} \tag{8.68}$$

and the configuration integral becomes

$$Z_\phi = V^N \left[1 - \frac{N^2 B(T)}{V} \right] \tag{8.69}$$

Here, $B(T)$ is called the second virial coefficient

$$B(T) = -2\pi \int_{-\infty}^{\infty} f(r) dr \tag{8.70}$$

If we were to drop the weakly interacting assumption and start admitting higher-order terms in Eq. (8.68), then we could more generally write

$$Z_\phi = V^N \left[1 - \frac{N^2 B(T)}{V} + \text{higher-order terms} \right] \tag{8.71}$$

8.6.2 Virial Equation of State

We saw that by noting

$$p = kT \frac{\partial \ln Q}{\partial V}\Bigg)_{1/T,N} \tag{8.72}$$

the only term in $\ln Q$ that contains V is $\ln Z_\phi$, so

$$p = kT \frac{\partial \ln Z_\phi}{\partial V}\Bigg)_{1/T,N} \tag{8.73}$$

Now, $\ln Z_\phi$ is in the form of $\ln 1 + x$, where x is small, and

$$\ln(1+x) = x - \frac{x^2}{2} + \cdots \tag{8.74}$$

Using this fact we were able to write p as

$$p = \frac{RT}{v} + \frac{a(T)}{v^2} + \frac{b(T)}{v^3} + \cdots \tag{8.75}$$

8.6.3 Other Properties

Given the partition functions, the other properties of interest can be calculated. Of particular interest is the internal energy. This becomes

$$U = \frac{3}{2}NkT + kT^2\left(\frac{\partial \ln Z_\phi}{\partial T}\right)_{NV} = \frac{3}{2}NkT + \bar{U} \tag{8.76}$$

where

$$\bar{U} = \frac{\int \cdots \int \phi e^{-\phi/kT} dr_1 dr_2 \ldots dr_N}{Z_\phi} \tag{8.77}$$

can be considered as the configuration energy.

8.6.4 Potential Energy Function

We discussed several potential energy functions used in thermodynamic analysis. These are illustrated in Fig. 8.1. Of these, the most important is the Lennard–Jones function

$$\phi(r) = 4\epsilon_{ij}\left[\left(\frac{\sigma_{ij}}{r}\right)^{12} - \left(\frac{\sigma_{ij}}{r}\right)^{6}\right] \tag{8.78}$$

Here, σ_{ij} is the interatomic distance at which the attractive and repulsive forces exactly balance and ϵ is the well depth where ϕ is a minimum. Lennard–Jones is commonly used, especially for calculating transport properties. However, it, and other more realistic potential functions, do not result in an algebraic form for $B(T)$, rather the integral expression

$$B(T) = 2\pi \int_0^\infty \left[1 - \exp\left\{-\frac{4\epsilon}{kT}\left[\left(\frac{\sigma}{r}\right)^{12} - \left(\frac{\sigma}{r}\right)^{6}\right]\right\}\right] r^2 dr \tag{8.79}$$

must be evaluated numerically. The integrals can often be evaluated in advance, as discussed in Section 8.4 and Appendix E.

8.6.5 Other Equations of State

We discussed two examples of equations of state that are empirical variations on the virial form. The Redlich–Kwong equation is given by

$$P = \frac{RT}{v-b} - \frac{a}{T^{1/2}v(v+b)} \tag{8.80}$$

The Benedict–Webb–Rubin equation of state is

$$P = \frac{RT}{v} + \frac{B_0RT - A_0 - C_0/T^2}{v^2} + \frac{bRT-a}{v^3} + \frac{a\alpha}{v^6} + \frac{c}{v^3T^2}\left(1+\frac{\gamma}{v^2}\right)\exp\left(\frac{-\gamma}{v^2}\right) \tag{8.81}$$

Parameters for this and other equations of state can be found in Cooper and Goldfrank [39], Perry and Green [8], and Hirschfelder, Curtiss, and Bird [14].

8.7 Problems

8.1 Expand the van der Waals equation of state in a virial expansion, and express the second virial coefficient in terms of the van der Waals constants a and b.

8.2 Using the Lennard–Jones potential, numerically calulate the second virial coefficient CO. Plot it as a function of temperature from 300 to 1000 K.

8.3 Evaluate the specific volume of CO_2 at 30 bar and 300 K by the

(a) Ideal gas law.

(b) Redlich–Kwong equation.
The Redlich–Kwong equation is given by

$$P = \frac{RT}{v-b} - \frac{a}{\sqrt{T}v(v+b)}$$

where

$$a = 0.42748\frac{R^2T_c^{2.5}}{P_c} \text{ and } b = 0.08664\frac{RT_c}{P_c}$$

(c) Compressibility chart.

9 Liquids

Liquids represent a significant complication over the systems we have discussed so far. In the case of dense gases, for example, to evaluate the configuration integral we assumed that the total potential energy is the sum of the pairwise potential energies between molecules. We pointed out that this only strictly applies to gases that are not too dense. However, in a liquid, atoms and/or molecules are very closely spaced, and the potential experienced by each particle depends on the positions of all the surrounding particles out to some distance. This is true in solids as well, but unlike solids, in a liquid there is diffusion and particles change position with respect to each other constantly.

As a result, statistical methods are most suited to analyzing liquid behavior. In this chapter we will introduce the concept of distribution functions and ultimately relate thermodynamic properties to the radial distribution function. Finally, we will discuss the use of molecular dynamics simulations to obtain the radial distribution function and the thermodynamic properties.

9.1 The Radial Distribution Function and Thermodynamic Properties

This presentation closely follows that in McQuarrie's excellent text [31]. We start by introducing the generalized distribution function, where Z_ϕ is the configuration integral:

$$P^{(N)}(r_1, r_2, \ldots, r_N) dr_1 dr_2 \cdots dr_N = \frac{e^{-\phi/k_B T} dr_1 dr_2 \cdots dr_N}{Z_\phi} \tag{9.1}$$

This function describes the probability that molecule 1 is in volume dr_1 at r_1, molecule 2 in dr_2 at r_2, and so on. However, N is normally a very large number. Suppose we deal with a reduced system with n molecules and ask what is the probability that molecule 1 is in dr_1 at $r_1, \ldots,$ molecule n is in dr_n at r_n, irrespective of the configuration of the remaining $N - n$ molecules. This can be written

$$P^{(n)}(r_1, r_2, \ldots, r_n) = \frac{\int \cdots \int e^{-\phi/k_B T} dr_{n+1} \cdots dr_N}{Z_\phi} \tag{9.2}$$

Assuming all the particles are identical, we could ask what is the probability that any n of them are located in position $r_1 \dots r_n$ irrespective of the configuration of the rest. This is the n-particle density, ρ in this development is a number density:

$$\rho^{(n)}(r_1, \dots, r_n) = \frac{N!}{(N-n)!} P^{(n)}(r_1, \dots, r_n) \tag{9.3}$$

The simplest distribution function is $\rho^{(1)}(r_1)$, which is the number density. In a crystal this will be a periodic function, but in a fluid all points are equivalent and $\rho^{(1)}(r_1)$ is independent of r_1. Therefore,

$$\frac{1}{V}\int \rho^{(1)}(r_1)dr_1 = \rho^{(1)} = \frac{N}{V} = \rho \tag{9.4}$$

We now define a correlation function, $g^{(n)}$, relating ρ^n to $\rho^{(n)}$:

$$\rho^{(n)}(r_1, \dots, r_n) = \rho^n g^{(n)}(r_1, \dots, r_n) \tag{9.5}$$

$g^{(n)}$ is called a correlation function since, if the molecules were independent of each other, $\rho^{(n)}$ would simply be ρ^n. Thus, $g^{(n)}$ corrects for non-independence or correlation.

One can show that

$$\begin{aligned} g^{(n)}(r_1, \dots, r_n) &= \frac{V^n N!}{N^n (N-n)!} \frac{\int \cdots \int e^{-\phi/k_B T} dr_{n+1} \cdots dr_N}{Z_\phi} \\ &= V^n(1 + \mathcal{O}(N^{-1})) \frac{\int \cdots \int e^{-\phi/k_B T} dr_{n+1} \cdots dr_N}{Z_\phi} \end{aligned} \tag{9.6}$$

$g^{(2)}(r_1, r_2)$ is especially important since it can be determined experimentally using light-scattering techniques. In a liquid of spherical symmetry, where particles are randomly distributed, $g^{(2)}$ depends only on the relative distance between particles. As a result, it is usually designated merely as $g(r)$ and called the "radial distribution function" (RDF).

The integral of the radial distribution function over all space is

$$\int_0^\infty \rho g(r) 4\pi r^2 dr = N - 1 \approx N (\text{for large } N) \tag{9.7}$$

Figures 9.1 and 9.2 illustrate the RDF idea. Figure 9.1 shows what the RDF would look like for a crystalline solid. Because the structure is ordered, the RDF is composed of a series of lines with regular spacing. For a liquid, however, the arrangement is random, and changing with time. The left side of Fig. 9.2 shows a typical random distribution of particles. Because geometry limits the location of nearby particles, the average density of particles as a function of radius is somewhat periodic, as shown in a typical liquid RDF plotted on the right. The RDF is zero at the origin because only the central particle can exist there. At large distances, the RDF goes to unity because the positions of other particles with respect to the target particle are completely randomized.

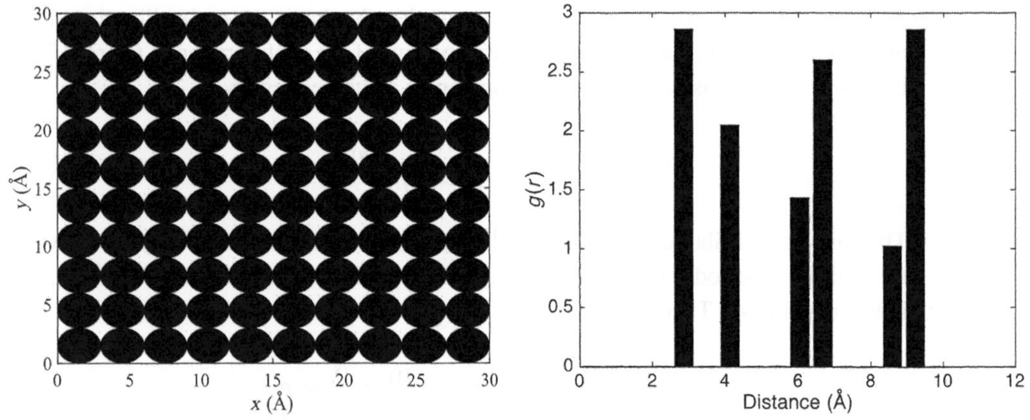

Figure 9.1 Radial distribution function for crystal structures.

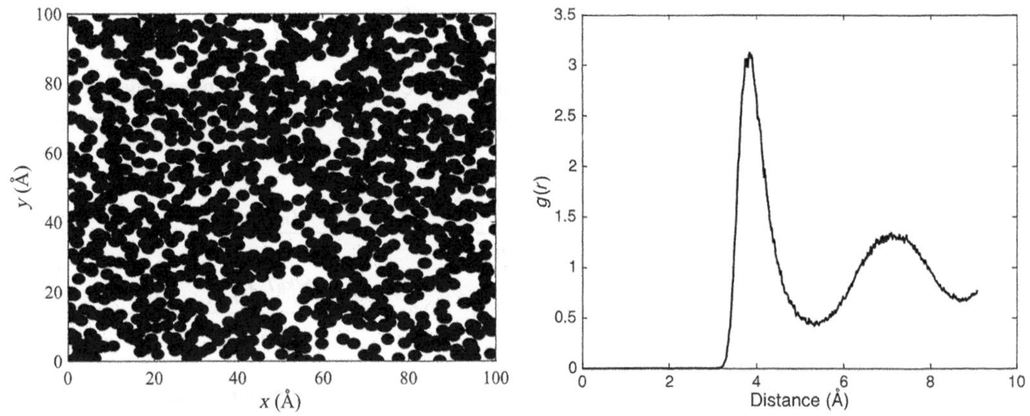

Figure 9.2 Radial distribution function for liquids.

One can show that if we assume that the total potential energy of the N-body system is pairwise additive, that is

$$U_N(r_1, \ldots, r_N) = \sum_{i<j} \phi(r_{ij}) \tag{9.8}$$

then all thermodynamic functions can be written in terms of $g(r)$.

The number density of molecules with centers between r and $r + dr$ relative to a specific molecule is

$$4\pi r^2 \frac{N}{V} g(r) dr \tag{9.9}$$

For a spherically symmetric pairwise potential function, it follows that

$$4\pi r^2 \phi(r) \frac{N}{V} g(r) dr \tag{9.10}$$

is the intermolecular potential energy between the central molecule and other molecules with centers between r and $r + dr$. One can then show that

$$u = \frac{3}{2}Rk_BT + \frac{N_A}{2}\rho \int_0^\infty \phi(r)g(r,\rho,T)4\pi r^2 dr \tag{9.11}$$

and

$$p = \rho k_BT - \frac{\rho^2}{6}\int_0^\infty r\frac{d\phi(r)}{dr}g(r,\rho,T)4\pi r^2 dr \tag{9.12}$$

9.1.1 Example 9.1

Calculate (a) the internal energy per mole and (b) the pressure for liquid argon using a radial distribution function calculated using molecular dynamics, assuming that van der Waals forces are calculated using the Lennard–Jones potential. Use a temperature of 135 K and a mass density of 1071.1 kg/m^3. (c) To check the results, compare with the values given in the NIST program REFPROP.

(a) The radial distribution function illustrated in Fig. 9.2 was calculated for these conditions. Using Lennard–Jones parameters of $\epsilon = 1.66 \times 10^{-21}$ J and $\sigma = 3.4 \times 10^{-10}$ m, the integral becomes (factoring out 4π and integrating using Matlab)

$$\int_0^\infty \phi(r)g(r,\rho,T)r^2 dr = -6.7465 \times 10^{-50} \text{ J m}^{-3}$$

The number density ρ_0 is equal to $\rho/m = 1.6151 \times 10^{28}$ m^{-3} and the configuration contribution to the internal energy is

$$u_{conf} = 2\pi\rho_0 N_A \int_0^\infty \phi(r)g(r,\rho,T)r^2 dr = -4519.8 \times 10^6 \text{ J/mol}$$

Adding in the ideal gas contribution

$$u_{id} = \frac{3}{2}R_uT = 1.783 \times 10^6 \text{ J/kmol}$$

the total internal energy becomes

$$u = -2.7365 \times 10^6 \text{ J/kmol}$$

(b) Using the same radial distribution function, and noting that

$$\frac{d\phi(r)}{dr} = 4\epsilon\left[\frac{\sigma^{12}}{r^{13}} - \frac{\sigma^6}{r^7}\right]$$

then

$$\int_0^\infty r\frac{d\phi(r)}{dr}g(r,\rho,T)4\pi r^2 dr = -8.4779 \times 10^{-49} \text{ J m}^3$$

Then

$$p_{conf} = -\frac{\rho_0^2}{6} \int_0^\infty r \frac{d\phi(r)}{dr} g(r, \rho, T) 4\pi r^2 dr = 3.6856 \text{ MPa}$$

The ideal gas contribution is

$$p_{id} = \rho_0 kT = 3.1887 \text{ MPa}$$

and the pressure becomes

$$p = 6.8743 \text{ MPa}$$

(c) REFPROP gives the following properties at 135 K and 1071.1 kg/m^3: $p = 6.3714$ MPa and $h = -58,158$ J/kg. Since $h = u - p/\rho$ and converting to kmols by multiplying by the molecular weight 39.948 kg/kmol, one obtains $u = -2.5609 \times 10^6$ J/kmol. Thus, both the pressure and internal energy are in reasonably close agreement to the values given in REFPROP, considering that the RDF came from a relatively small molecular dynamics simulation.

9.2 Molecular Dynamics Simulations of Liquids

Molecular dynamics (MD) involves following the microscopic behavior at the atomistic level by making the Born–Oppenheimer approximation and solving $F = ma$ for all the nuclei present. One can calculate the forces using either empirical or quantum methods, but empirical methods are almost always used for reasons of computational efficiency. The advantage of MD is that it accounts directly for the thermal motion of atoms and molecules. It directly outputs position and velocity of all particles as a function of time. Thus, it provides all information about phase space, the space determined by the position, velocity, and orientation of all molecules. If phase space is known at all times, then both the thermodynamic properties and dynamic behavior of the system can be determined. MD has been used for a wide array of problems, including protein folding, thermodynamic properties, stress–strain relations, surface tension of drops, behavior of supercritical fluids, micro heat transfer, and diffusional processes.

MD solvers start with Newton's Law

$$\vec{F} = m\vec{a} \tag{9.13}$$

where the force is usually provided in terms of the potential function experienced by each nuclei:

$$\vec{F}_i = -\frac{\partial V}{\partial \vec{r}_i} \tag{9.14}$$

It is usual to rewrite Newton's Law as

$$\vec{a}_i = \frac{\vec{F}_i}{m_i} \tag{9.15}$$

for each particle i, resulting in a set of second-order ordinary differential equations that must be solved simultaneously:

$$\frac{d^2\vec{x}_i}{dx^2} = \frac{\vec{F}_i}{m_i} \tag{9.16}$$

A number of solution techniques have been used in MD simulations. However, because realistic simulations are often quite large, efforts have been made to develop very fast solvers at the expense of some accuracy. One of the most common is the Verlet method, which is developed by writing Taylor series expansions for the position forward and backward in time:

$$\begin{aligned} \vec{r}(t+\Delta t) &= \vec{r}(t) + \vec{v}(t)\Delta t + \frac{1}{2}\vec{a}(t)\Delta t^2 + \frac{1}{6}\vec{b}(t)\Delta t^3 + \mathcal{O}(\Delta t^4) \\ \vec{r}(t-\Delta t) &= \vec{r}(t) - \vec{v}(t)\Delta t + \frac{1}{2}\vec{a}(t)\Delta t^2 - \frac{1}{6}\vec{b}(t)\Delta t^3 + \mathcal{O}(\Delta t^4) \end{aligned} \tag{9.17}$$

where a is the acceleration and b the third derivative of position with time. Adding the two expressions, one obtains

$$\vec{r}(t+\Delta t) = 2\vec{r}(t) - \vec{r}(t-\Delta t) + \vec{a}(t)\Delta t^2 + \mathcal{O}(\Delta t^4) \tag{9.18}$$

The difficulty with this method is that the velocities are calculated. One could use

$$\vec{v}(t) = \frac{\vec{r}(t+\Delta t) - \vec{r}(t-\Delta t)}{2\Delta t} \tag{9.19}$$

However, this is only second-order accurate in time. A better scheme is to do the following, which is fourth-order accurate in both position and velocity:

$$\begin{aligned} r(t+\Delta t) &= r(t) + v(t)\Delta t + \frac{1}{2}a(t)\Delta t^2 \\ v(t+\Delta t/2) &= v(t) + \frac{1}{2}a(t)\Delta t \\ a(t+\Delta t) &= \frac{F(r(t+\Delta t))}{m} \\ v(t+\Delta t) &= v(t+\Delta t/2) + \frac{1}{2}a(t+\Delta t)\Delta t \end{aligned} \tag{9.20}$$

Note how we need $9N$ memory locations to save the $3N$ positions, velocities, and accelerations, but we never need to have simultaneously stored the values at two different times for any one of these quantities. An alternative method used by some programs is the leapfrog Verlet, which uses a time-splitting approach:

$$\begin{aligned} v(t+\Delta t/2) &= v(t-\Delta t/2) + a(t)\Delta t \\ v(t) &= \frac{1}{2}\left(v\left(t+\frac{1}{2}\Delta t\right) + v\left(t-\frac{1}{2}\Delta t\right)\right) \\ r(t+\Delta t) &= r(t) + v(t)\left(t+\frac{1}{2}\Delta t\right)\Delta t \end{aligned} \tag{9.21}$$

One can, of course, use more accurate solvers at the expense of greater computational resource requirements. Examples include Beeman's method, Runge–Kutta, and Gera's method.

To carry out the simulation requires knowledge of the potential energy function. Equation (9.22) is in the so-called Amber form:

$$V = \sum_{Bonds} K_r(r - r_{eq})^2 + \sum_{Angles} K_\theta(\theta - \theta_{eq})^2 + \sum_{Dihedrals} V_n(1 + \cos(n\phi - \gamma)) + \sum_{i=1}^{N-1}\sum_{j>i}^{N}\left[\frac{A_{ij}}{R_{ij}^{12}} - \frac{B_{ij}}{R_{ij}^{6}} + \frac{q_i q_j}{r_{ij}}\right] \tag{9.22}$$

The first term is harmonic bond stretching, the second three-atom bond bending, and the third four-atom torsion. The final term includes non-bonded Lennard–Jones attraction and repulsion and electrostatic interaction. A_{ij} and B_{ij} are written in terms of the well depth and size parameter ϵ and σ:

$$\begin{aligned} A_{ij} &= 4\varepsilon_{ij}\sigma_{ij}^{12} \\ B_{ij} &= 4\varepsilon_{ij}\sigma_{ij}^{6} \end{aligned} \tag{9.23}$$

The Lennard–Jones parameters are obtained from the Lorentz–Berthelot mixing rule:

$$\begin{aligned} \varepsilon_{ij} &= \sqrt{\varepsilon_i \varepsilon_j} \\ \sigma_{ij} &= (\sigma_{ii} + \sigma_{jj})/2 \end{aligned} \tag{9.24}$$

There are numerous details in implementing MD solvers that are beyond the scope of this book. Several excellent books on the subject are listed in the References.

9.3 Determining $g(r)$ from Molecular Dynamics Simulations

$g(r)$ can be determined experimentally. However, it is quite easy to obtain from molecular dynamics. For a particular type of atom at a particular time,

$$g(r) = \frac{\langle N(r, \Delta r)\rangle}{\frac{N}{2}\rho V(r, \Delta r)} \tag{9.25}$$

where the numerator is the average over all such atoms. $N(r, \Delta r)$ is the number of atoms found in a spherical shell between r and $r + \Delta r$, $V(t, \Delta r)$ is the volume of the spherical shell, N the total number of atoms sampled, and ρ the total density (number per unit volume) within the sampled volume.

If the simulation is run for M time steps, then the time average is

$$g(r) = \frac{\sum_{k=1}^{M} \langle N(r, \Delta r)\rangle}{M\frac{N}{2}\rho V(r, \Delta r)} \tag{9.26}$$

As an example, consider a simulation of 150 Ar atoms at conditions corresponding to the liquid state. The simulation was run for 100 ps, with an integration step size of 1 fs, and output every 1 ps. The results are those shown in Fig. 9.2.

9.4 Molecular Dynamics Software

There are a number of open source and commercial classical MD packages that are suitable for calculating the properties of liquids. (By classical we mean that the equations of motion for the individual particles are derived from Newton's Law.) Among the best known open source programs are LAMMPS, DLPoly, and Tinker. LAMMPS [40] (http://lammps.sandia.gov) is provided by Sandia National Laboratories and has a number of features that make it useful for large-scale calculations. It can deal with a large number of particle and model types and force fields, and a variety of container configurations. The current version is written in C++ and comes with a set of utilities written in Python scripting language. DL_POLY [41] (https://www.scd.stfc.ac.uk/Pages/DL_POLY.aspx) is a general-purpose MD simulation package developed by I. T. Todorov and W. Smith at the UK Research Council Daresbury Laboratory. Written in Fortran90, it is scalable and can be run on a single processor to high-performance machines. Tinker [42] (https://dasher.wustl.edu/tinker/) is another general-purpose MD program in Fortran. A new version designed for massively parallel systems has also just been released. Tinker comes with a number of useful utilities.

Each of these programs has its charms, and which you will prefer depends on the user's particular needs and programming language preference (if you intend to make modifications). All are adaptable to very large systems and have been used to study gas, liquid, and solid states. Only LAMMPS provides the user with a ReaxFF force field, which allows the modeling of chemical reactions using classical molecular mechanics.

9.4.1 Example 9.2

Use the Tinker MD program to calculate the radial distribution function of water at room temperature and pressure using the waterbig example.

Download the complete distribution at https://dasher.wustl.edu/tinker/. You can also get executables for Linux, MacOS, and/or Windows, which avoids having to build the executables. For this example you will need three executables: dynamic, archive, and radial. Dynamic is the MD program that solves the equations of motion. It produces configuration files (with xyz extensions) at a sequence of times from a starting configuration. Archive allows you to consolidate xyz files into a single binary file. Radial is used to calculate the radial distribution function after running a simulation.

To start you need to select a force field, set up a key file that contains run instructions, and provide the initial configuration of molecules in the form of an xyz file. In this example, because water is a simple molecule, the force-field parameters can be included in the key file. You will find key and xyz files for waterbig in the example directory.

The key file is listed here. The force-field parameters come from the TIP3P force field. The meaning of each parameter in the key file is listed in more detail in the users' manual.

```
verbose
neighbor-list
a-axis     36.342
ewald
rattle     WATER
tau-temperature     1.0
vdwtype                          LENNARD-JONES
radiusrule                       GEOMETRIC
radiustype                       SIGMA
radiussize                       DIAMETER
epsilonrule                      GEOMETRIC
dielectric                         1.0
atom           1    O              "O Water (TIP3P)"      8      15.999     2
atom           2    H              "H Water (TIP3P)"      1       1.008     1
vdw            1                     3.150656111      0.152072595
vdw            2                     0.000        0.000
bond         1    2              529.6         0.9572
angle          2    1     2      34.05         104.52
ureybrad   2     1    2              38.25         1.5139
charge         1                      -0.834
charge         2                       0.417
```

The example directory also contains the starting configuration file, the first four lines of which are shown below. The first line is for comments. In this case there are 4800 atoms, the simulation takes place in a 36.342^3 Å^3 box, and the force field is TIP3P. The next three lines give the index number, the atom name, the x, y, z coordinates, and the connectivity of the first water molecule. The remainder of the file contains the coordinates and connectivity of the remaining water molecules. You should make a copy of this file, waterbig_init.xyz, for example. Then make a copy as waterbig.xyz for input to dynamic, as running dynamic will change the file and you may want to run more than one simulation:

```
4800 Water Cubic Box (36.342 Ang, 1600 TIP3P)
1 O 7.801866 15.203752 -3.825662 1 2 3
2 H 7.593339 15.479832 -4.718147 2 1
3 H 7.125906 14.561598 -3.608981 2 1
```

A reasonable sequence of commands is

```
cp waterbig_init.xyz waterbig.xyz
./dynamic waterbig 100000 2.0 1.0 2 298.15 > waterbig.out
./archive waterbig 1 1 200 1
./radial waterbig 1 200 1 1 1 0.01 n > waterbig_rdf.txt
rm waterbig.xyz*
rm waterbig.dyn
rm waterbig.0*
rm waterbig.1*
```

The first command is to copy waterbig_init.xyz to waterbig.xyz.

Dynamic is then run with the following parameters:

100,000	number of time steps
2.0	time step (fsec)
1.0	time between data dumps
2	ensemble type, canonical in this case
298.15	simulation temperature

The parameters for radial are:

waterbig	archived file name
1	initial time step
200	last time step
1	incremental time step
1 1	1st and 2nd atom type names
0.01	width of averaging pins
n	do not include intramolecular pairs

This simulation on a Macbook Pro took approximately 1 h.

This is followed by archiving the data starting at step 1, in increments of 1 step, out to 200 steps. Next, radial is run to calculate the radial distribution function as in Eq. (9.26) using the same steps that have been archived and then writing the output to a text file, waterbig_rdf.txt. Finally, unneeded files are deleted.

Once the simulation is complete, you can visualize the time evolution of the simulation using the open source program Force Field Explorer that comes with Tinker. An example image of one time step is shown in Fig. 9.3.

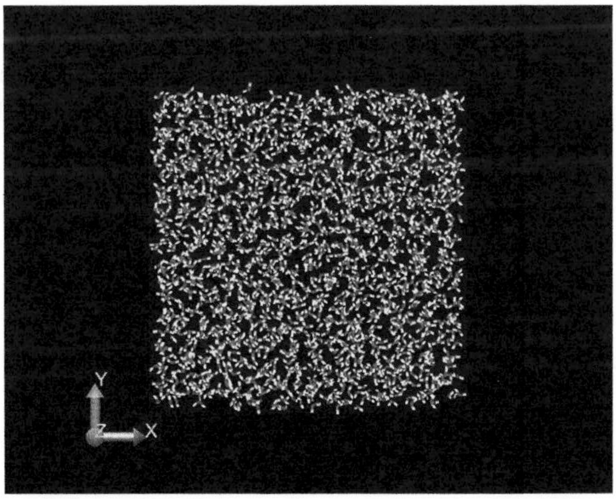

Figure 9.3 Water molecules in simulation box.

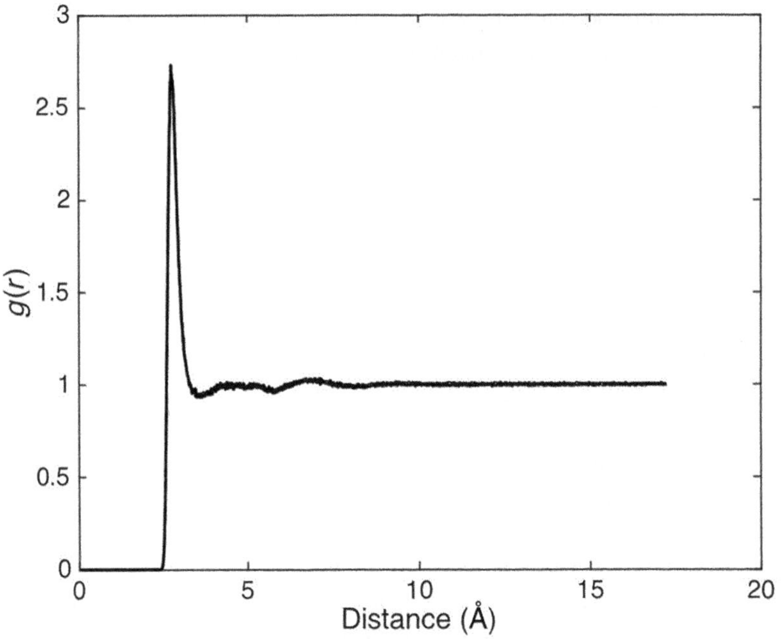

Figure 9.4 Radial distribution function.

Finally, any plotting program can display the radial distribution function using the data in waterbig_rdf.txt (Fig. 9.4).

9.5 Summary

In this chapter we discussed the use of distribution functions to describe liquid behavior. In particular, the radial distribution function (RDF), $g(r)$, plays an important role. We showed that the internal energy and pvT relations can be obtained from the RDF:

$$u = \frac{3}{2}Rk_BT + \frac{N_A}{2}\rho \int_0^\infty \phi(r)g(r,\rho,T)4\pi r^2 dr \tag{9.27}$$

and

$$p = \rho k_BT - \frac{\rho^2}{6}\int_0^\infty r\frac{d\phi(r)}{dr}g(r,\rho,T)4\pi r^2 dr \tag{9.28}$$

We then explored the use of molecular dynamics simulations to determine $g(r)$ and other properties.

9.6 Problems

Radial Distribution Function

9.1 The purpose of this exercise is to explore the role of MD simulation in determining the properties of liquids, argon in this case. You may use any MD program you like, there are many open source codes available. Your task is to determine the density as a function of temperature and pressure, and then the radial distribution function. You will use the RDF to calculate the internal energy using Eq. (9.11). You will compare how two force fields, Lennard–Jones and Buckingham potential, affect behavior.

(a) An experimental paper by Itterbeek and Verbeke [43] reports measurements of the density of liquid nitrogen and argon as a function of temperature and pressure. The data are for the "compressed liquid" domain. You will see that for a given temperature, the pressure is a very strong function of the density. If you use an NPT ensemble, the density becomes a dependent variable and an output of the simulation. Select 89.13 K in Table II of Itterbeek and Verbeke and then vary the pressure over the values in the table (use seven equally spaced values from the table). Set the temperature and pressure using a thermostat and a barostat. For each pressure, determine the density from the simulation.

You will notice that the density fluctuates, so calculate the standard deviation too. One question in MD simulations is when to start averaging and how many time steps to average over. If you start too soon, then the result may be biased by the transient startup. One way to decide when you have waited long enough to start averaging is to plot both the average and the standard deviation as a function of time. Once the numbers have settled down, then start averaging. If you don't average over enough time steps then the variance, or standard deviation, will be too large. Once you have decided that you have good numbers, then plot the density along with the data in Table II.

(b) After you have run each *T–P* simulation, calculate the radial distribution function. Then write a script in your favorite programming environment (Matlab, Mathematica, etc.) to perform the integration in Eq. (9.11). Compare the results with the energy output by the MD program.

10 Crystalline Solids

Solids are materials in which intermolecular forces hold the atoms and molecules relatively rigidly in place. They generally fall into two classes, crystalline and amorphous. Crystalline solids have a regular, ordered rigid structure. They have precise melting and boiling points. Examples are most metals, diamond, sugar, and graphite. Amorphous solids don't have a regular geometric shape. Their structure varies randomly in three-dimensional space and they melt over a range of temperatures. Examples are coal, plastic, and rubber. The thermodynamic treatment of these two forms is very different. As a result, amorphous solids are typically treated as a separate subject which is beyond the scope of this book. On the other hand, crystalline solids are relatively easy to treat, as we shall see below.

If the solid has a crystalline structure then the concept of normal modes of vibration can be used to treat the crystal as a set of independent particles. One can show that the thermodynamic properties are a function of the vibrational spectrum of the crystal. We start by exploring the nature of the vibrational spectrum and two theories, Einstein and Debye, that make simple assumptions about the spectra.

A crystal can, in many cases, be represented as a set of atoms connected by springs, as shown in two dimensions in Fig. 10.1. The springs represent the interatomic forces holding the atoms in place. Each atom is located in a potential well, as shown in Fig. 10.2. If the atomic displacement is small, the force can be represented as harmonic, that is

$$\vec{F} = -k\vec{x} \tag{10.1}$$

where $\vec{x}$ is the atomic displacement and k a spring constant. Thus,

$$V(x) = k(x - x_0)^2 \tag{10.2}$$

The development to obtain the partition function is similar to that for a dense gas. However, there are no modes of motion other than vibration. The total energy can be written as

$$U = U_0 + \phi(\vec{r_1}, \vec{r_2}, \ldots) \tag{10.3}$$

where U_0 represents the zero-point energy for the crystal as a whole. The natural frequency of a harmonic oscillator is

$$\nu_j = \frac{1}{2\pi}\left(\frac{k_j}{\mu_j}\right)^{1/2} \tag{10.4}$$

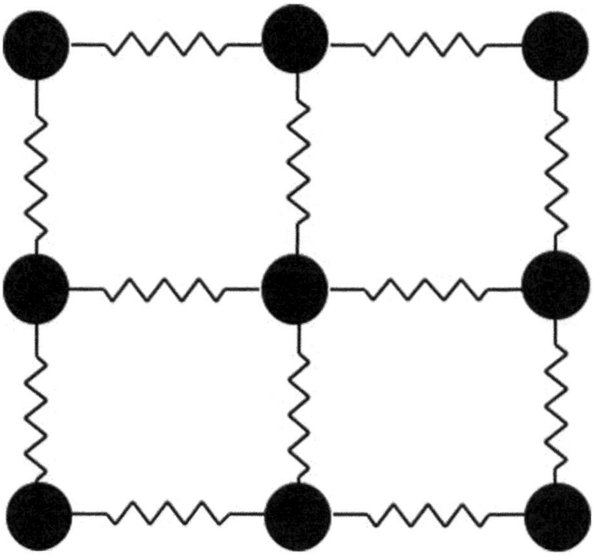

Figure 10.1 Crystal structure in 2D.

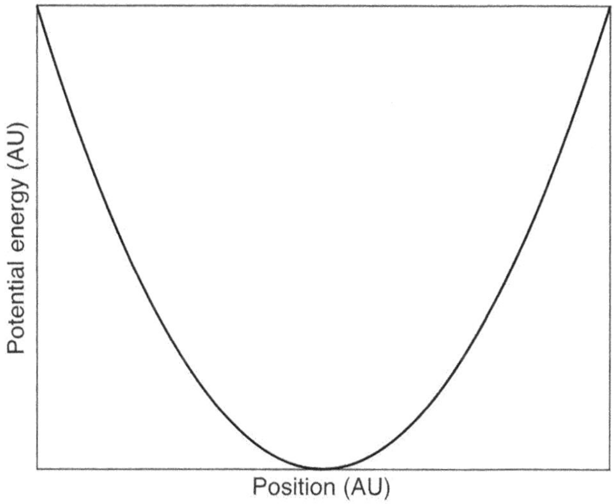

Figure 10.2 The dependence of average potential energy (per atom) on position.

where k_j and μ_j are an effective force constant and an effective reduced mass. Their values will depend on the details of the crystal structure and there will be a distribution of frequencies based on the normal modes.

If we assume that the normal modes of motion are independent of each other, then the molecular partition function for vibration is the product of the individual normal mode partition functions:

$$q_{vib} = \prod_{j=1}^{3N-6} q_{vib,j} \tag{10.5}$$

Taking the zero-point energy into account, the canonical partition function becomes

$$Q = e^{-\frac{U_0}{kT}} \prod_{j=1}^{3N-6} q_{vib,j} \tag{10.6}$$

We have derived the molecular partition function for a harmonic oscillator in our examination of ideal gases:

$$q_{vib} = \frac{e^{-h\nu/2kT}}{1 - e^{-h\nu/kT}} \tag{10.7}$$

Thus, the total partition function becomes

$$Q = e^{-\frac{U_0}{kT}} \prod_{j=1}^{3N-6} \left(\frac{e^{-h\nu_j/2kT}}{1 - e^{-h\nu_j/kT}} \right) \tag{10.8}$$

Taking the negative logarithm of Q (remember that $S[1/T] = -k \ln Q$), we have

$$-\ln Q = \frac{U_0}{kT} + \sum_{j=1}^{3N} \frac{e^{-h\nu_j/2kT}}{1 - e^{-h\nu_j/kT}} = \frac{U_0}{kT} + \sum_{j=1}^{3N} \left[\ln (1 - e^{-h\nu_j/kT}) + \frac{h\nu_j}{2kT} \right] \tag{10.9}$$

In a macroscopic crystal, the normal-mode frequencies are essentially continuously distributed. If we introduce a distribution function $g(\nu)$, and convert the sum to an integral, we get

$$-\ln Q = \frac{U_0}{kT} + \int_0^\infty \left[\ln (1 - e^{-h\nu/kT}) + \frac{h\nu}{2kT} \right] g(\nu) d\nu \tag{10.10}$$

Here, $g(\nu)$ is normalized to the total number of normal-mode frequencies

$$\int_0^\infty g(\nu) d\nu = 3N \tag{10.11}$$

We can now calculate the thermodynamic properties. The Helmholtz function is just

$$F[1/T] = -\ln Q = \frac{U_0}{kT} + \int_0^\infty \left[\ln (1 - e^{-h\nu/kT}) + \frac{h\nu}{2kT} \right] g(\nu) d\nu \tag{10.12}$$

Then the internal energy is

$$U = U_0 + \int_0^\infty \left[\frac{h\nu e^{-h\nu/kT}}{1 - e^{-h\nu/kT}} + \frac{h\nu}{2} \right] g(\nu) d\nu \tag{10.13}$$

and the specific heat becomes

$$c_v = k \int_0^\infty \left[\frac{(h\nu/kT)^2 e^{-h\nu/kT}}{(1 - e^{-h\nu/kT})^2} \right] g(\nu) d\nu \tag{10.14}$$

It remains to determine $g(\nu)$.

10.1 Einstein Crystal

Einstein made the very simple assumption that there was only a single normal-mode frequency. If we define the characteristic Einstein temperature as

$$\Theta_E \equiv \frac{h\nu_{vib}}{k} \tag{10.15}$$

then the Helmholtz function, the energy, and the specific heat become

$$\frac{F}{kT} = -\left\{\frac{NU_0}{2kT} + 3N\ln\left(2\sinh\frac{\theta_E}{2T}\right)\right\} \tag{10.16}$$

$$U = \frac{NU_0}{2} + \frac{3}{2}Nk\theta_E\coth\frac{\theta_E}{2T} \tag{10.17}$$

$$C_v = 3Nk\left[\frac{\theta_E/2T}{\sinh(\theta_E/2T)}\right]^2 \tag{10.18}$$

The entropy is

$$S = 3Nk\left[\frac{\theta_E}{2T}\coth\frac{\theta_E}{2T} - \ln\left(2\sinh\frac{\theta_E}{2T}\right)\right] \tag{10.19}$$

The specific entropy, energy, and specific heat are plotted in Fig. 10.3 as a function of T/θ_E.

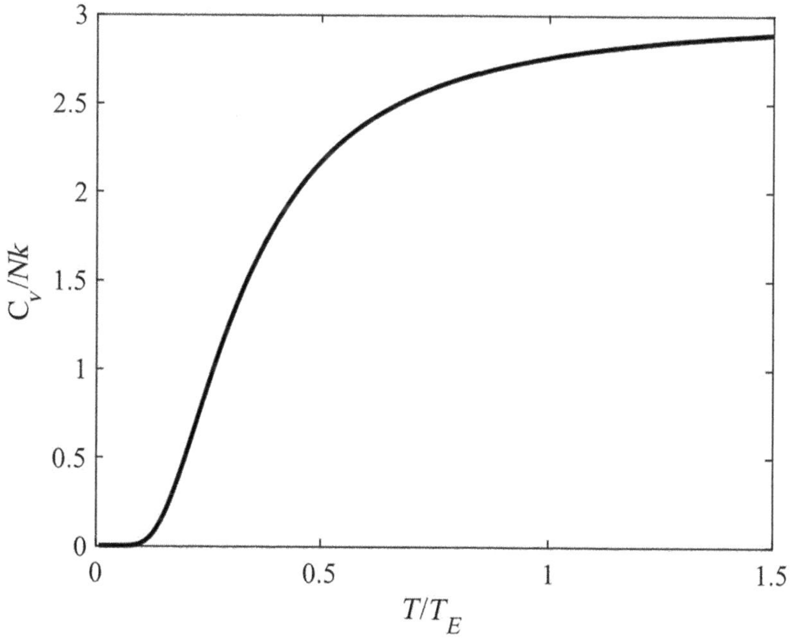

Figure 10.3 Properties of the Einstein crystal.

The limiting values of C_v are

$$
\begin{aligned}
T \to 0, \quad & \frac{C_v}{Nk} \to 3\left(\frac{\theta_E/T}{e^{\theta_E/2T}}\right)^2 \\
T \to \infty, \quad & \frac{C_v}{Nk} \to 3
\end{aligned}
\tag{10.20}
$$

10.2 Debye Crystal

The Einstein solid works pretty well, except that interparticle interactions are ignored. Every oscillator is the same. In fact, at significant temperatures, a solid acts more like a multibody system with $3N-6$ degrees of freedom. Debye assumed that he could determine the distribution using elastic wave analysis and ignoring quantum effects. The wave analysis results in a wave velocity distribution that is essentially the same as the translational energy distribution from the particle in a box problem. The bottom line is that

$$
g(\nu)d\nu = \left(\frac{2}{\mathrm{v}_t^3} + \frac{1}{\mathrm{v}_l^3}\right) 4\pi V \nu^2 d\nu \tag{10.21}
$$

where v_t and v_l are the transverse and longitudinal wave speeds in the solid. Defining an effective wave speed as

$$
\frac{3}{\mathrm{v}_0^3} = \frac{2}{\mathrm{v}_t^3} + \frac{1}{\mathrm{v}_l^3} \tag{10.22}
$$

$g(\nu)$ becomes

$$
g(\nu)d\nu = \frac{12\pi V}{\mathrm{v}_0^3} \nu^2 d\nu \tag{10.23}
$$

Because of the normalization of $g(\nu)$, there is a maximum allowed frequency. This is

$$
\nu_m = \left(\frac{3N}{4\pi V}\right)^{1/3} \mathrm{v}_0 \tag{10.24}
$$

so

$$
\begin{aligned}
g(\nu)d\nu &= \left(\frac{9N}{\nu_m^3}\right) \nu^2 d\nu, \quad 0 \le \nu \le \nu_m \\
g(\nu)d\nu &= 0, \quad \nu > \nu_m
\end{aligned}
\tag{10.25}
$$

ν_m is called the "cutoff" frequency.

The difference between the Einstein and Debye frequency distributions is illustrated in Fig. 10.4. In effect, Eistein assumed that $g(\nu)$ was a single-frequency delta function.

Plugging all this into the fundamental relation, we obtain

$$
\frac{F}{kT} = -N\left\{\frac{U_0}{2kT} + \frac{9}{8}\frac{\theta_D}{T} + 3\ln\left(1 - e^{-\theta_D/T}\right) - 3\left(\frac{T}{\theta_D}\right)^3 \int_0^{\theta_D/T} \frac{x^3 dx}{e^x - 1}\right\} \tag{10.26}
$$

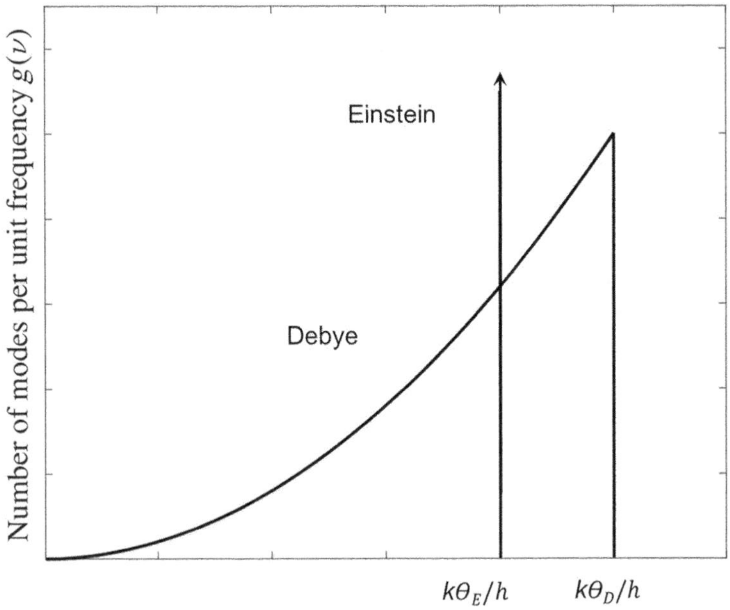

Figure 10.4 Frequency distributions for Einstein and Debye models.

where

$$\theta_D \equiv \frac{h\nu_m}{k} \tag{10.27}$$

is called the Debye temperature. Then

$$\frac{U}{kT} = -N\left\{\frac{U_0}{2kT} + \frac{9}{8}\frac{\theta_D}{T} + 9\left(\frac{T}{\theta_D}\right)^3 \int_0^{\theta_D/T} \frac{x^3 dx}{e^x - 1}\right\} \tag{10.28}$$

$$\frac{S}{k} = 9N\left(\frac{T}{\theta_D}\right)^3 \int_0^{\theta_D/T} \left[\frac{x^3}{e^x - 1} - x^2 \ln\left(1 - e^{-x}\right)\right] dx \tag{10.29}$$

$$\frac{C_v}{k} = 3ND\left(\frac{T}{\theta_D}\right) \tag{10.30}$$

where the "Debye function" is defined as

$$D\left(\frac{T}{\theta_D}\right) \equiv 3\left(\frac{T}{\theta_D}\right)^3 \int_0^{\theta_D/T} \frac{x^4 e^x}{(e^x - 1)^2} dx \tag{10.31}$$

The limiting values of C_v for the Debye crystal are

$$\begin{aligned} T \to 0, &\quad \frac{C_v}{Nk} \to \frac{12\pi^4}{5}\left(\frac{T}{\theta_D}\right)^2 \\ T \to \infty, &\quad \frac{C_v}{Nk} \to 3 \end{aligned} \tag{10.32}$$

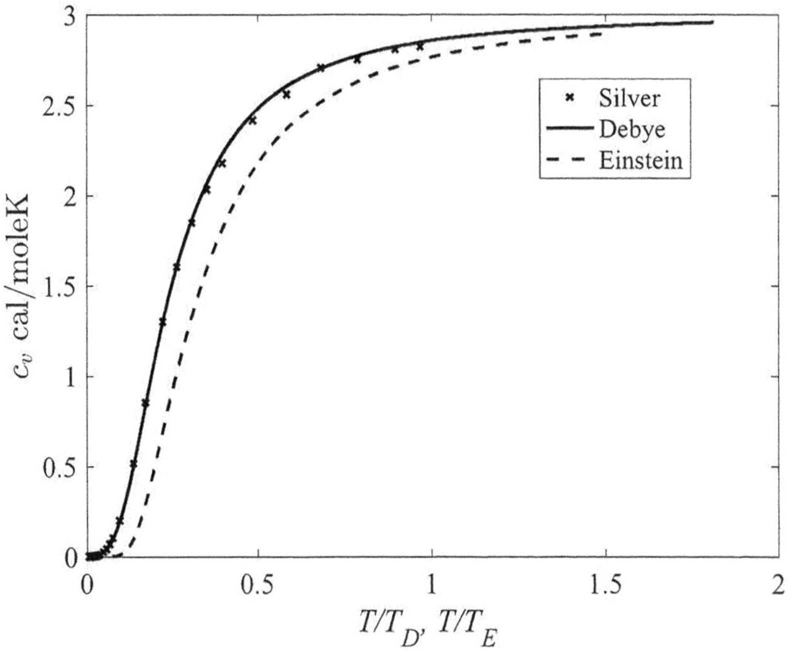

Figure 10.5 Specific heat for Einstein and Debye models (data from Kittel [10] and White and Collocott [11]).

The specific heat for both the Einstein and Debye models is shown in Fig. 10.5, along with data for crystalline silver. Although both models predict the same value at high temperatures, at lower temperatures the Debye model provides a better fit.

10.2.1 Example 10.1

(a) The specific heat of copper is known to be 4.70 cal/gmol K at 150 K. Using this value, calculate the Einstein temperature of copper.

(b) Calculate the specific heat of the copper at 50 K using both the Einstein and Debye theories. (The Debye temperature of copper is 315 K.)

(c) Calculate the electron contribution to the specific heat at 50 K.

(a) This problem can be solved easily using Matlab or Mathematica. To find the Einstein temperature, find the zero of

$$0 = C_v - 3Nk\left[\frac{\theta_E/2T}{\sinh(\theta_E/2T)}\right]^2$$

The result is 256.5 K.

(b) The Einstein specific heat is calculated using Eq. (10.34). To calculate the Debye specific heat requires the Debye function, Eq. (10.37), which can easily be calculated using Matlab or Mathematica. Then, the Debye specific heat is

$$\frac{C_{vD}}{k} = 3ND\left(\frac{T}{\theta_D}\right)$$

The numerical calculations result in

$$\frac{C_{vE}}{Nk} = 3.93 \text{ J/mol K}$$

$$\frac{C_{vD}}{Nk} = 5.994 \text{ J/mol K}$$

(c) The electron contribution to the specific heat is given by [Eq. (7.42)]

$$C_v = \frac{\pi^2}{2} N_e k \frac{T}{T_F}$$

The Fermi level and free electron number density for copper are given in Table 7.1 as $\mu_0 = 7.00$ eV and $N_e = 8.47 \times 10^{22}$ cm^{-3}. Therefore, the Fermi temperature is 81,202 K and

$$C_v = 0.0036 \text{ J/mol K}$$

Thus, the electronic contribution to the specific heat is quite small.

Both the Einstein and Debye temperatures can be inferred from specific heat data or from elastic constants. The Debye temperature tends to decrease with increasing atomic mass. This is because the frequency of a harmonic oscillator is proportional to $\sqrt{k/m}$. Typical values of θ_D are fairly low, usually of the order of room temperature or less. This is due to the relatively low force constants and large atomic masses in monatomic crystals. In general, Debye theory fits fairly well for monatomic crystal solids. It does not work for non-crystal solids, which are beyond the scope of this book. Table 10.1 lists Debye temperatures for a number of materials.

Table 10.1 Debye temperatures for some monatomic crystalline solids

Solid	θ_D(K)	Solid	θ_D(K)
Na	150	Fe	420
K	100	Co	385
Cu	315	Ni	375
Ag	215	Al	390
Au	170	Ge	290
Be	1000	Sn	260
Mg	290	Pb	88
Zn	250	Pt	225
Cd	172	C (diamond)	1860

10.3 Summary

In this chapter we discussed two relatively simple theories for crystalline solids. From a thermodynamic viewpoint the main property of interest is the specific heat. We showed that, in general, the specific heat can be written as

$$c_v = k \int_0^\infty \left[\frac{(h\nu/kT)^2 e^{-h\nu/kT}}{(1 - e^{-h\nu/kT})^2} \right] g(\nu) d\nu \tag{10.33}$$

where $g(\nu)$ is a vibrational distribution function. The difference between the two theories lies in how they calculate $g(\nu)$. The Einstein theory assumes there is only a single vibrational frequency. The resulting expression for the specific heat is

$$C_{v_E} = 3Nk \left[\frac{\theta_E/2T}{\sinh(\theta_E/2T)} \right]^2 \tag{10.34}$$

where the characteristic Einstein temperature is

$$\theta_E \equiv \frac{h\nu_{vib}}{k} \tag{10.35}$$

In the Debye theory, a distribution of frequencies is assumed based on a normal-mode analysis. In this case the specific heat becomes

$$\frac{C_{v_D}}{k} = 3ND\left(\frac{T}{\theta_D}\right) \tag{10.36}$$

where the "Debye function" is defined as

$$D\left(\frac{T}{\theta_D}\right) \equiv 3\left(\frac{T}{\theta_D}\right)^3 \int_0^{\theta_D/T} \frac{x^4 e^x}{(e^x - 1)^2} dx \tag{10.37}$$

where

$$\theta_D \equiv \frac{h\nu_m}{k} \tag{10.38}$$

is the Debye temperature.

10.4 Problems

10.1 A very simple assumption for the specific heat of a crystalline solid is that each vibrational mode of the solid acts independently and is fully excited and thus $c_v = 3N_A k_B = 24.9$ kJ/kmol K. This is called the law of Dulong and Petit. Compare this result with the Debye specific heat of diamond at room temperature, 298 K. Use a Debye temperature of 2219 K.

10.2 Using the experimental value for the specific heat of gold at 155 K of $C_v/R = 2.743$, estimate the Einstein characteristic temperature. (MW gold = 197.0.)

10.3 A 1-kg slab of Al initially at 300 K is quenched by placing it in a bath of liquid nitrogen (77 K). What is the heat loss to the nitrogen for this process, assuming the Al can be described using the Debye model? Assume that $c_v = 797$ J/kg K at 200 K to determine the appropriate Debye temperature to use.

10.4 (a) Calculate the specific heat of magnesium at 50 K using the Debye theory.
(b) Calculate the electron contribution to the specific heat.

11 Thermodynamic Stability and Phase Change

Phase change is interesting because it presents the possibility of two or more phases coexisting with the same pressure and temperature. In this chapter we explore the role of thermodynamic stability in determining the allowable phases of a substance. The presentation follows Callen [3] closely.

11.1 Thermodynamic Stability

Recall the fundamental problem of thermodynamics, as outlined in Chapters 1 and 2. To find the equilibrium state following the removal of a constraint we found the maximum of the entropy by setting its derivative equal to zero, $dS = 0$, and solving for the independent variables that satisfied that condition. However, finding a zero derivative of a function only means that there is a critical point, namely a minimum, $d^2S \geq 0$, a maximum, $d^2S \leq 0$, or a saddle point, $d^2S = 0$. Here we explore the consequences of those cases where the second derivative is not less than zero.

Start by considering Fig. 11.1. Assume that the two sides contain the same substance and have the same fundamental relations but are initially isolated from one another. Then, for the system as a whole,

$$S_{\text{initial}} = S_1(U_1, V_1, N_1) + S_2(U_2, V_2, N_2) \tag{11.1}$$

Now suppose that we explore transferring a small amount of energy from subsystem 1 to subsystem 2. The final value of the entropy for the system would be

$$S_{\text{final}} = S_1(U_1 - \Delta U, V_1, N_1) + S_2(U_2 + \Delta U, V_2, N_2) \tag{11.2}$$

The value of the second derivative will depend on the shape of $S(U)$. If the initial state is located in a concave region of $S(U)$, then $d^2S > 0$ and the state is not located at a maximum of S. This is illustrated in Fig. 11.1, which shows that $S_{\text{final}} > S_{\text{initial}}$. If the barrier to heat transfer were removed, then energy would flow between the two subsystems. Hence, the initial condition is unstable.

This example illustrates the concept of mutual stability, that is the stability of two or more systems with respect to each other. This is what we explored in Chapter 2, where we determined that for two systems to be in equilibrium with respect to each other requires that their intensive properties T, P, and μ be equal. Intrinsic stability, which is

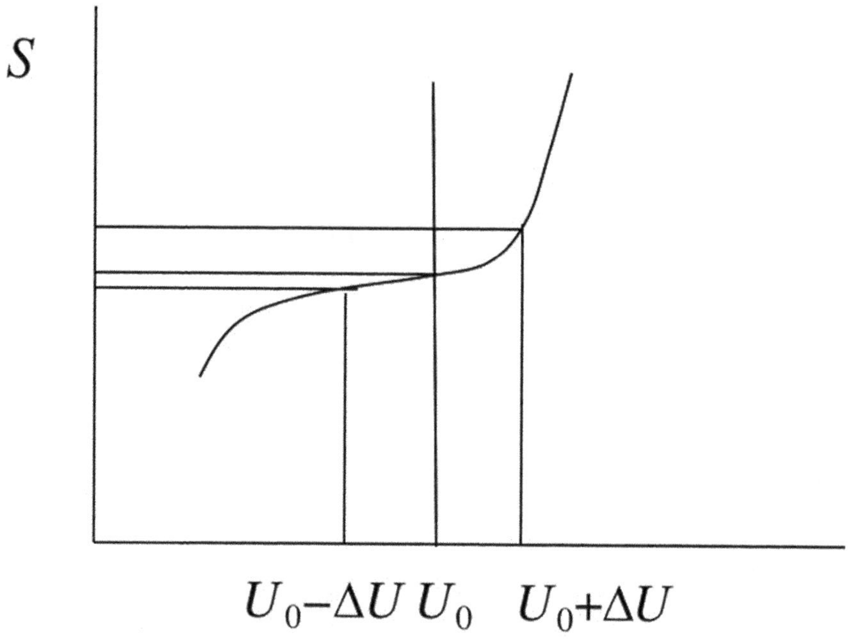

Figure 11.1 The fundamental relation.

our interest in this chapter, refers to the stability of a single system, and to phase stability in particular.

Suppose we have a single system in an initial thermodynamic state. Imagine that in a very small volume, dV, the energy is perturbed a small amount, dU. By conservation of energy, the energy of the remaining system must change by an amount $-dU$. Thus (ignoring changes in V and N),

$$dS = S_{ss}(U_{ss} + dU) + S_s(U_s - dU) \tag{11.3}$$

where the subscripts ss refer to the subsystem and s to the remainder. If dS is greater than zero, then the initial state must be unstable. Such can be the case if the fundamental relation has the shape illustrated in Fig. 11.1.

It is possible that a fundamental equation shows the behavior illustrated in Fig. 11.2. Although the regions BC and EF appear to be stable, in practice the effective fundamental relation will be given by the tangent line E′. The two regions BC and EF are said to be locally stable, while the region BCDEF is globally unstable. Points on the tangent line E′ have separated phases.

While we have illustrated the stability problem by postulating small changes in energy, the same reasoning applies to perturbation in the volume or mole numbers. If $f = f(x, y)$, and f_x and f_y are the first derivatives of f with respect to x and y, respectively, then the total second derivative of the function is

$$D(x, y) = f_{xx}f_{yy} - f_{xy}^2 \tag{11.4}$$

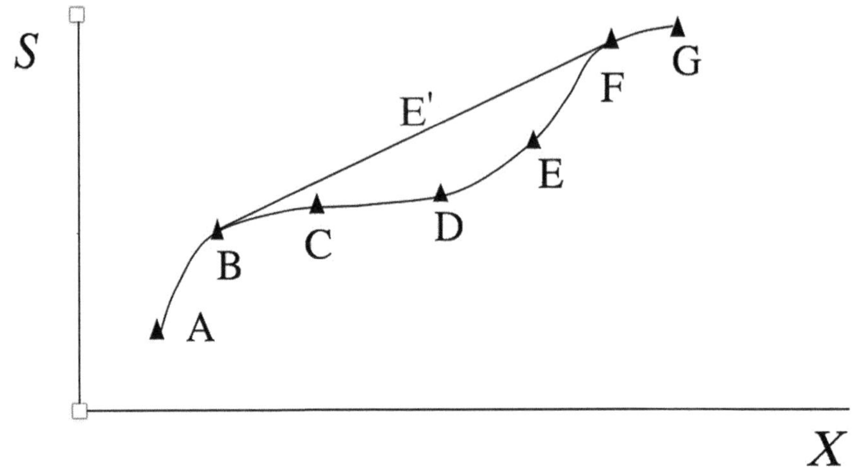

Figure 11.2 Example fundamental relation.

where $f_{xx} = \partial^2 f/\partial x \partial x$, and so on. This leads to the following possibilities:

- If $D > 0$ and $f_{xx} < 0$, then $f(x, y)$ has a relative maximum at (x, y).
- If $D > 0$ and $f_{xx} > 0$, then $f(x, y)$ has a relative minimum at (x, y).
- If $D < 0$, then $f(x, y)$ has a saddle point at (x, y),
- If $D = 0$, the second derivative test is inconclusive.

Recall that while the entropy and its transforms, the Massieu functions, are maximized, the energy and the thermodynamic potentials are minimized. For example, this requires that for the potentials, the stability criteria are

$$\frac{\partial^2 U}{\partial S^2}\frac{\partial^2 U}{\partial V^2} - \left(\frac{\partial^2 U}{\partial S \partial V}\right) \geq 0 \tag{11.5}$$

$$\left.\frac{\partial^2 F}{\partial T^2}\right)_{V,N} \leq 0 \text{ and } \left.\frac{\partial^2 F}{\partial V^2}\right)_{T,N} \geq 0 \tag{11.6}$$

$$\left.\frac{\partial^2 H}{\partial T^2}\right)_{P,N} \geq 0 \text{ and } \left.\frac{\partial^2 H}{\partial P^2}\right)_{S,N} \leq 0 \tag{11.7}$$

$$\left.\frac{\partial^2 G}{\partial T^2}\right)_{P,N} \leq 0 \text{ and } \left.\frac{\partial^2 G}{\partial P^2}\right)_{T,N} \leq 0 \tag{11.8}$$

Using Maxwell's relations, one can show that local stability also requires

$$\kappa_T \geq \kappa_S \geq 0 \tag{11.9}$$

$$C_p \geq C_v \geq 0 \tag{11.10}$$

$$\left.\frac{\partial P}{\partial v}\right)_T = -\frac{1}{\kappa_T} \leq 0 \tag{11.11}$$

where κ_T and κ_S are the isothermal and isentropic compressibility, respectively.

11.2 Phase Change

Now consider the stability of a substance described by van der Waals equation:

$$P = \frac{RT}{v - b} - \frac{a}{v^2} \tag{11.12}$$

Local stability requires that

$$\left.\frac{\partial P}{\partial v}\right)_T \leq 0 \tag{11.13}$$

If we plotted p–v isotherms of van der Waals equation in the liquid vapor region it would look something like Fig. 11.3. This is typical of many simple compressible substances. A single isotherm is shown in Fig. 11.4. The local stability criterion is clearly violated over the region F–K–M and therefore a phase transition must be involved.

We can explore this more exactly by calculating the Gibbs potential, noting that the chemical potential is the Gibbs function per mole in the Gibbs representation and stable points have minimum μ:

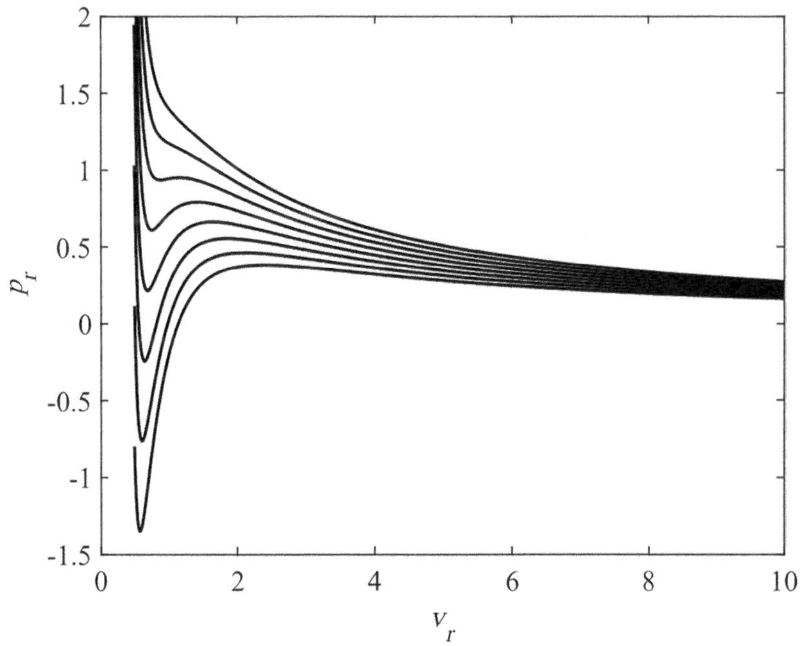

Figure 11.3 Isotherms of typical pvT relationship.

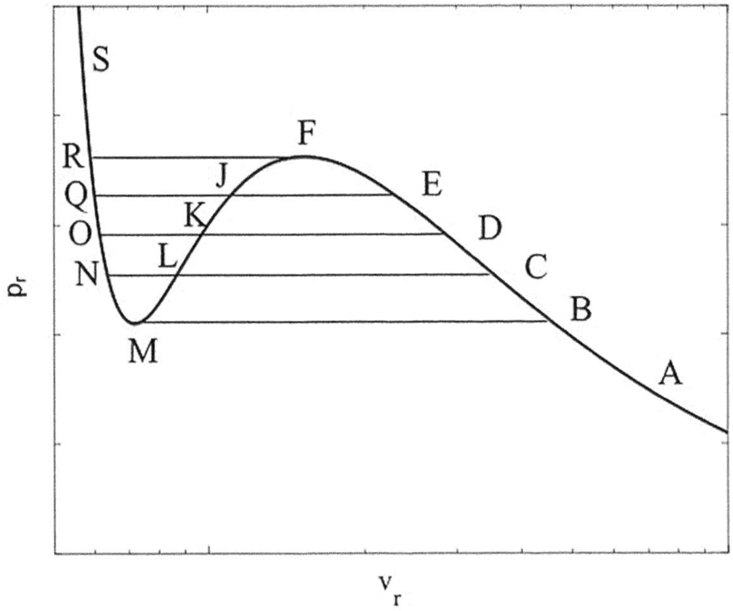

Figure 11.4 Single isotherms of typical pvT relationship.

$$d\mu = -sdT + vdp \tag{11.14}$$

Along a line of constant T,

$$\mu = \int vdp + \phi(T) \tag{11.15}$$

where $\phi(T)$ is an integration constant (that will be a function of T). Thus, the change in chemical potential between two points on an isotherm is

$$\mu_A - \mu_B = \int_A^B v(p)dp \tag{11.16}$$

If we were to carry out the integration over a range of pressures, the results would look like Fig. 11.5.

As the pressure is increased, the physical state is the one with minimum μ. At the points D, O where the two sides overlap, both liquid and vapor branches are at minimum μ. Recall the pvT phase diagram for a simple compressible substance (Fig. 11.6). The phase change lines are locations where two branches of the Gibbs potential meet and μ is at a minimum. As a result, the pv behavior of an isotherm must look something like shown in Fig. 11.7, where the isotherm falls on the line DKO and not DFKMO. The areas I and II must be equal, because

$$\mu_D - \mu_O = \int_O^D v(p)dp = 0 \tag{11.17}$$

Finally, the vapor dome is constructed by applying Eq. (11.17) over the full range of volumes.

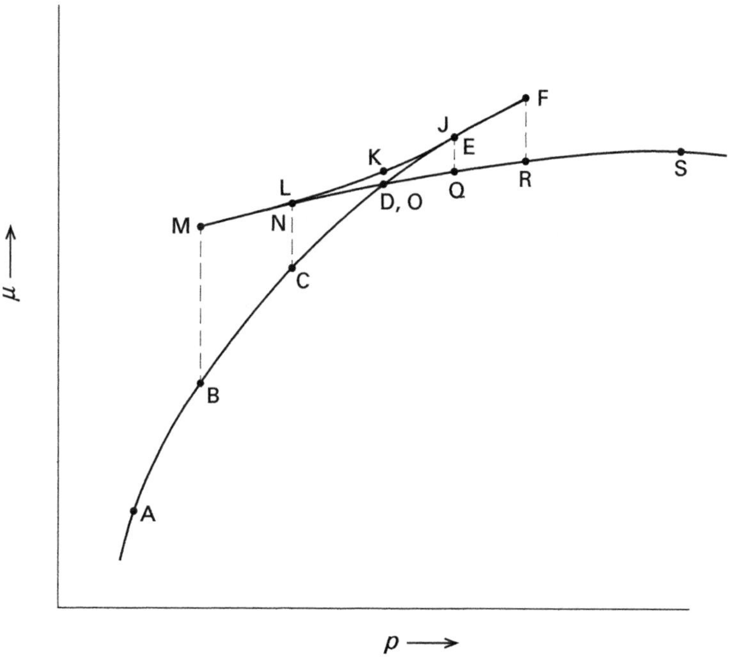

Figure 11.5 Gibbs potential as a function of p.

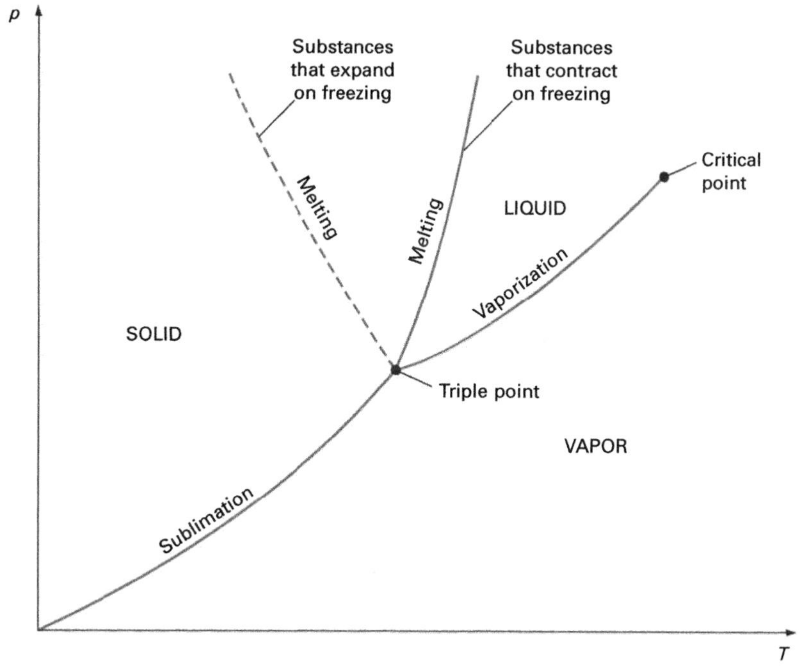

Figure 11.6 Phase diagram for simple compressible substance.

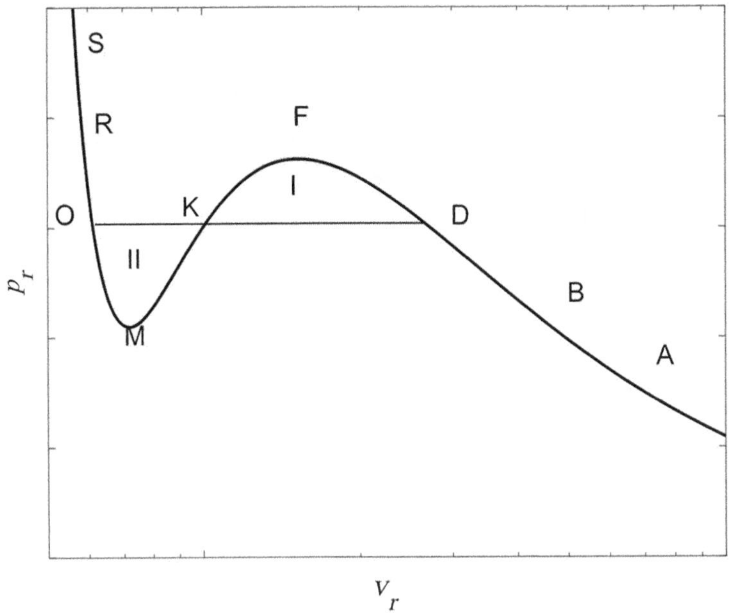

Figure 11.7 Isotherm of typical pv diagram.

11.2.1 Example 11.1

A particular substance satisfies the van der Waals equation of state. Plot the p–v relation as a function of reduced (p/p_c, etc.) pressure and volume for a temperature of 0.95 T_c. Then find the saturation region and obtain the reduced p_{sat}, v_f, and v_g.

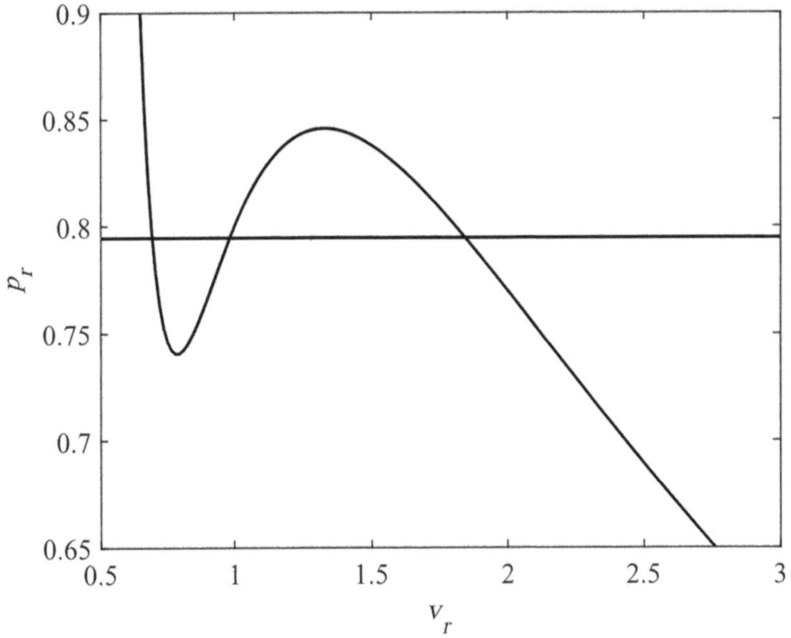

Figure 11.8 Van der Waals equation.

This problem is best solved numerically. Plot the van der Waals equation and search for the reduced saturation pressure for which the areas are equal. In doing so, one obtains $v_f = 0.69$ and $v_g = 1.84$.

11.2.2 Example 11.2

Two gram moles of the van der Waals fluid are maintained at a temperature of 0.95 T_c and a volume of 200 cm^3. Find the quality of the mixture. Use the critical parameters for water and the values of v_{g_r} and v_{f_r} from Example 11.1.

For H_2O, $T_c = 647.1$ K, $p_c = 22.06 \times 10^6$ Pa, and $v_c = 0.056$ m^3/kmol.

Given $N_{moles} = 2 \times 10^{-3}$ kmols, $T_r = 0.95$, and $V = 200 \times 10^{-6}$ m^3, then $v = V/N_{moles} = 0.002$ m^3/kmols, $v_f = v_{f_r}v_c = 0.0386$ m^3/kmols, $v_g = v_{g_c}v_c = 0.103$ m^3/kmols, and the quality

$$x = \frac{v - v_f}{v_g - v_f} = 0.9528$$

The saturation properties can be calculated using

$$s_{fg} = s_D - s_O = \int_{OMKFD} \left(\frac{\partial p}{\partial T}\right)_v dv \tag{11.18}$$

$$u_{fg} = u_D - u_O = Ts_{fg} - p(v_D - v_O) \tag{11.19}$$

$$h_{fg} = (u_D + p_D v_D) - (u_O + p_O v_O) = u_{fg} + p(v_D - v_O) \tag{11.20}$$

One can easily derive the Clapeyron equation by writing s as a function of v and T and taking the derivative

$$ds = \left(\frac{\partial s}{\partial v}\right) dv + \left(\frac{\partial s}{\partial T}\right) dT \tag{11.21}$$

During a phase change the temperature is constant, so

$$ds = \left(\frac{\partial s}{\partial v}\right) dv \tag{11.22}$$

Using Maxwell's relation

$$ds = \left(\frac{\partial p}{\partial T}\right)_v dv \tag{11.23}$$

Since the pressure and temperature are constant, the derivative cannot change, and can be replaced with the total derivative:

$$ds = \frac{dp}{dT} dv \tag{11.24}$$

Integrating from one phase to the other,

$$\frac{dp}{dT} = \frac{\Delta s}{\Delta v} \tag{11.25}$$

Now

$$du = Tds - pdv \tag{11.26}$$

Introducing the enthalpy,

$$dh = du + pdv = Tds \tag{11.27}$$

Given constant pressure and temperature,

$$\Delta s = \frac{\Delta h}{T} = \frac{L}{T} \tag{11.28}$$

where L is the latent heat phase change. Therefore,

$$\frac{dp}{dT} = \frac{L}{T\Delta v} \tag{11.29}$$

11.3 Gibbs Phase Rule

The stability requirement that the Gibbs free energy be minimized also applies in the case of multicomponent mixtures. The complexity grows with the number of components. However, there is a simple rule, called the Gibbs phase rule. If r is the number of components, and M the number of phases, then

$$f = r - M + 2 \tag{11.30}$$

where f is the number of independent intensive parameters allowed. f is also called the *thermodynamic degrees of freedom*. The number of phases that can coexist is $M = f + r + 2$. The maximum number of phases that can coexist is found by setting $f = 0$. For a single-component system, a maximum of three phases can coexist. We call that point the triple point, as shown in Fig. 11.6. The possibilities for a two-component system are illustrated in Table 11.1.

Table 11.1 Number of phases for a two-component system

f	M
0	4
1	3
2	2
3	1

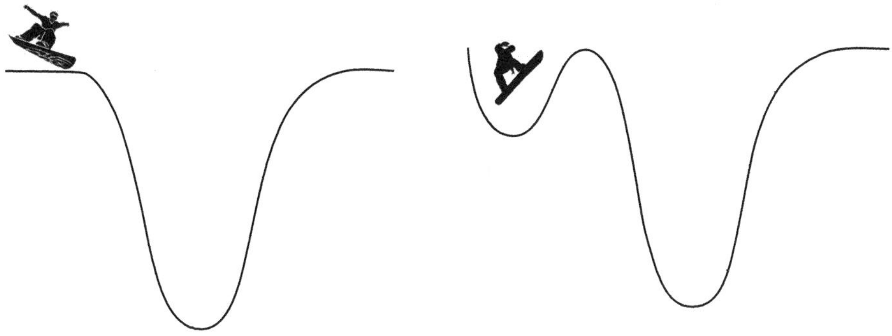

Figure 11.9 Global versus local stability.

11.4 Thermodynamic versus Dynamic Stability

Just because the thermodynamics suggest that a phase change should occur, in practice it might not. This is because action at the microscopic level is required for the phase change to occur. A classic example is the heat treating of steel, where quenching "locks in" a harder phase state than would exist at room temperature were the metal to be at equilibrium. Another example is the supersaturation of dissolved sugar in water.

To understand this phenomenon, consider the two snowboarding halfpipes (potential energy curves) illustrated in Fig. 11.9. Imagine that a snowboarder is poised to drop into the pipe (well), as shown on the left side. For the snowboarder it is all downhill, no work is required to "fall" into the well. The snowboarder on the right, however, finds herself in a small hole on the lip of the potential. To get into the halfpipe (well) will require work to get over the hump. In thermodynamic systems that little bit of work must come from fluctuating motions at the microscopic level. If they are large enough, then the system can be pushed over into a globally stable configuration.

The snowboarder on the right is in a local energy minimum, and therefore a "local" stable point, while the bottom or the large well is a "global" stability point.

11.5 Summary

11.5.1 Thermodynamic Stability

In this section we discussed examining the second derivatives of the entropy and its various transformations. Whether a given equilibrium point is stable to perturbations in the independent parameters depends on the sign of the derivatives. This criterion can be written as

$$d^2S \geq 0 \tag{11.31}$$

and the potentials

$$\frac{\partial^2 U}{\partial S^2}\frac{\partial^2 U}{\partial V^2} - \left(\frac{\partial^2 U}{\partial S \partial V}\right) \geq 0 \tag{11.32}$$

$$\left.\frac{\partial^2 F}{\partial T^2}\right)_{V,N} \leq 0 \text{ and } \left.\frac{\partial^2 F}{\partial V^2}\right)_{T,N} \geq 0 \tag{11.33}$$

$$\left.\frac{\partial^2 H}{\partial T^2}\right)_{P,N} \geq 0 \text{ and } \left.\frac{\partial^2 H}{\partial P^2}\right)_{S,N} \leq 0 \tag{11.34}$$

$$\left.\frac{\partial^2 G}{\partial T^2}\right)_{P,N} \leq 0 \text{ and } \left.\frac{\partial^2 G}{\partial P^2}\right)_{T,N} \leq 0 \tag{11.35}$$

Using Maxwell's relations, one can show that local stability also requires

$$\kappa_T \geq \kappa_S \geq 0 \tag{11.36}$$

$$C_p \geq C_v \geq 0 \tag{11.37}$$

$$\left.\frac{\partial P}{\partial v}\right)_T = -\frac{1}{\kappa_T} \leq 0 \tag{11.38}$$

where κ_T and κ_S are the isothermal and isentropic compressibility, respectively.

11.5.2 Phase Change

We explored the issue of stability for van der Waals equation. We showed how the stability criterion

$$\left.\frac{\partial P}{\partial v}\right)_T \leq 0 \tag{11.39}$$

could be used to determine the location of the liquid–vapor saturation region, or vapor dome, as illustrated in Fig. 11.7.

We also derived expressions for s_{jg}, u_{fg}, and h_{fg}, along with the Clapeyron relation

$$\frac{dp}{dT} = \frac{L}{T\Delta v} \tag{11.40}$$

11.5.3 Gibb's Phase Rule

The stability requirement that the Gibbs free energy be minimized also applies in the case of multicomponent mixtures. The complexity grows with the number of components. However, there is a simple rule, called the Gibbs phase rule. If r is the number of components, and M the number of phases, then

$$f = r - M + 2 \tag{11.41}$$

where f is the number of independent intensive parameters allowed.

11.6 Problems

11.1 Consider the Redlich–Kwong equation of state:

$$P = \frac{RT}{v-b} - \frac{a}{T^{1/2}v(v+b)}$$

For methane, the parameters a and b are $a = 32.19$ bar(m^3/kg mol)^2 K$^{0.5}$ and $b = 0.02969$ m^3/kg mol.

Find the critical point predicted by this equation and compare the critical values to those in a standard thermodynamics book like Cengel and Boles [44].

11.2 A particular substance satisfies the Peng–Robinson equation of state. Plot the P–v relation as a function of reduced (p/p_c, etc.) pressure and volume for a temperature of 0.8 T_c. Then find the saturation region and obtain the reduced P_{sat}, v_f, and v_g.

11.3 Two gram moles of the Peng–Robinson fluid are maintained at a temperature of 0.8 T_c and a volume of 200 cm^3. Find the quality of the mixture. Use the critical parameters for water and the values of v_{g_r} and v_{f_r} from the previous problem.

12 Kinetic Theory of Gases

So far everything we have studied has involved systems at equilibrium. However, there are many non-equilibrium phenomena in the physical world, in particular those that occur in fluid flow. These include momentum, energy, and mass transfer with respect to the mean flow, as well as chemical non-equilibrium due to chemical reaction rates being slow compared to characteristic flow or diffusion times. In equilibrium the various modes of energy storage are distributed with the Boltzmann distribution (which for velocity becomes the Maxwellian). Because molecular–molecular or radiation–molecular interactions are required to maintain equilibrium, if characteristic flow times are short enough then the energy distributions and/or species concentrations can depart from their equilibrium values.

The conservation equations for fluid flow can be derived from a molecular perspective and we have done so in Appendix D. They take the following form.

Continuity:

$$\frac{\partial \rho}{\partial t} + \frac{\partial \rho v_i}{\partial x_i} = 0 \tag{12.1}$$

Species:

$$\frac{\partial \rho_k}{\partial t} + \frac{\partial \rho_k (v_i + v_{k_i})}{\partial x_i} = w_k \tag{12.2}$$

Momentum:

$$\frac{\partial \rho v_i}{\partial t} + \frac{\partial \rho v_i v_j}{\partial x_i} = -\frac{\partial p}{\partial x_i} + \frac{\partial \tau_{ij}}{\partial x_j} + \sum_{k=1}^{N} \rho_k \overline{F}_{k_i} \tag{12.3}$$

Thermal energy:

$$\frac{\partial u}{\partial t} + \rho v_j \frac{\partial u}{\partial x_j} = -\frac{\partial q_i}{\partial x_i} + \sigma_{ij} \frac{\partial v_i}{\partial x_j} + \dot{Q} + v_i \sum_{k=1}^{N} \rho_k \overline{F}_{k_i} \tag{12.4}$$

The most common type of non-equilibrium behavior involves the diffusional transport of momentum, energy, and/or mass that comes about due to gradients in velocity, temperature, or species concentration in continuum flows. These are accounted for by the shear stress tensor τ_{ij}, the heat flux vector q_i, and the species diffusion velocities v_{ki}. In the simple form you are probably most familiar with, these transport terms are typically written as

$$\tau = \mu \frac{du}{dy} \tag{12.5}$$

$$q = -\lambda \frac{dT}{dy} \tag{12.6}$$

$$J_1 = \rho_1 v_{1y} = -D_{12} \frac{dn_1}{dy} \tag{12.7}$$

where μ is the dynamic viscosity, λ the thermal conductivity, and D the binary diffusion coefficient.

The importance of the various terms in the conservation equations can be assessed through the use of several non-dimensional property groupings. These include the Reynolds, Prandtl, Schmidt, and Lewis numbers. Another important set of non-dimensional ratios are the Damkohler numbers that relate a chemical reaction rate to other characteristic rates, including convection and diffusion. These numbers are defined in Table 12.1.

An important parameter in the study of non-equilibrium flows is the Knudsen number

$$Kn = \frac{\lambda}{L} \tag{12.8}$$

where λ is the mean free path between collisions and L is a character length for the flow. If the mean free path is small compared to L, then the flow is described as collision dominated. This is the domain of continuous flow described by the Navier–Stokes or Euler equations. As the Knudsen number increases, the applicability of the continuum conservation equations begins to break down. Non-continuum effects are first felt at boundaries, and slowly extend into the body of the flow as the Knudsen number increases. At a large enough Knudsen number very few collisions occur. This is called the collisionless domain. These ideas are illustrated in Fig. 12.1.

Regardless of the flow domain, if the velocity distribution function is known, all the properties of the flow can be calculated. The time and spatial evolution of the velocity distribution function under all conditions is described by the Boltzmann equation. However, as we shall see, its solution is not straightforward. Therefore, we will start by examining a very simple approach to estimating transport properties. Then we will

Table 12.1 Non-dimensional parameters of fluid flow

Name	Value
Reynolds number	$Re = U * L/\nu$
Prandtl number	$Pr = \nu/\alpha$
Schmidt number	$Sc = \nu/D$
Lewis number	$Le = D/\alpha$
Damkohler number type I	$Da = \frac{\text{reaction rate}}{\text{convective flow rate}}$
Damkohler number type II	$Da = \frac{\text{reaction rate}}{\text{mass diffusion rate}}$

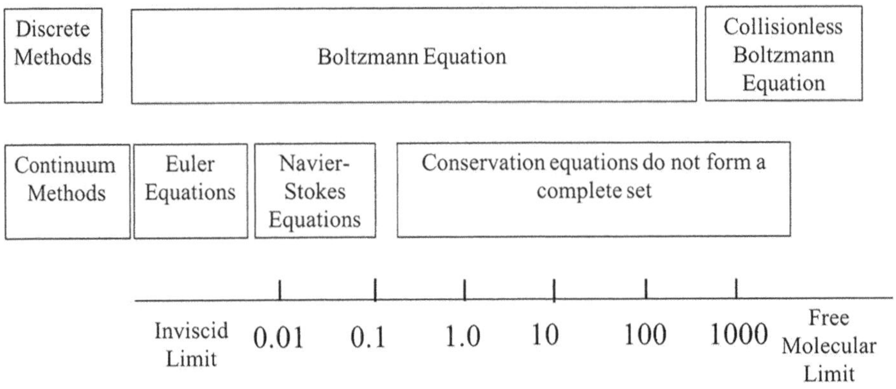

Figure 12.1 Relationship between Knudsen number and mathematical models (adapted from Bird [12]).

present the more exact theoretically derived expressions, and finally discuss practical approaches to obtaining the transport properties.

12.1 Transport Phenomena

Viscosity, thermal conductivity, and mass diffusivities are coefficients in relationships that describe the transport of momentum, energy, and mass when gradients in those properties are present. The physics of transport are as follows. Molecules are constantly moving, and as they move they carry properties with them including mass, momentum, and energy. While mass is preserved by individual atoms and molecules, their energy and momentum are determined by the history of collisions they have experienced. At equilibrium, there will be no net macroscopic flow of atomic and molecular properties. However, if there is a gradient of a macroscopic property, then the average flux of that property will not be the same in the positive and negative directions along the gradient.

12.1.1 Simple Estimates of Transport Rates

We can illustrate this concept using Fig. 12.2. Let ϕ be some property. The left plot shows how it varies in the y direction. Consider a plane located at $y = y_0$. Molecules that cross that plane carry, on average, a value of ϕ approximately corresponding to where their last collisions occurred. We have illustrated that location as a plane a distance δy above or below y_0. Because of the gradient in ϕ, the upward flux of ϕ will be greater than the downward flux, assuming that the number flux of molecules is approximately equal in the two directions.

The values of the property carried in the positive and negative directions are respectively

$$\phi(y_0 - \delta y) \text{ and } \phi(y_0 + \delta y) \tag{12.9}$$

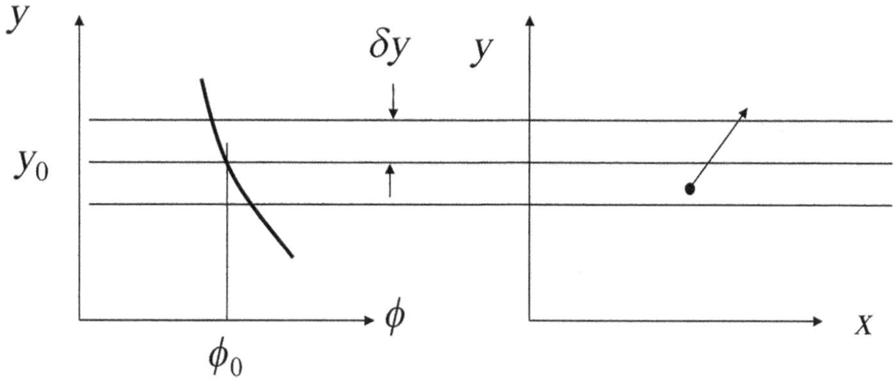

Figure 12.2 Gradient transport.

To determine the property flux we need to know the particle flux. In general, the flux of particles can be written as

$$J_i = \int c_i f(c_i) dc_i \tag{12.10}$$

where c_i is the component of particle velocity in the ith direction and $f(c_i)$ is the particle speed distribution in the given direction. The flux of some property ϕ carried in the same direction is thus

$$J_i = \int \phi c_i f(c_i) dc_i \tag{12.11}$$

where ϕ is evaluated at $\pm\delta y$ as appropriate.

We derived the Maxwellian speed distribution function in Chapter 5:

$$f(c_i) = \left(\frac{m}{2\pi kT}\right)^{1/2} \exp\left[-\frac{m}{2kT}c_i^2\right] \tag{12.12}$$

If this distribution function is substituted into Eq. (12.11) and the integration carried out we obtain (where n is the number density)

$$J_{\text{particles}} = \frac{n\bar{c}}{4} \tag{12.13}$$

as the number flux of particles across any plane. If we assume that even under non-equilibrium conditions with property gradients the speed distribution is approximately Maxwellian (and this holds well in the continuum regions at reasonably low macroscopic velocities), then we can use this expression to estimate the directional flux of the property of interest. In other words,

$$J_\phi = \phi\frac{n\bar{c}}{4} \tag{12.14}$$

where as before

$$\bar{c} = \sqrt{\frac{8kT}{\pi m}} \tag{12.15}$$

We next need to consider what value to use for δy. δy is the characteristic distance traveled on average by molecules since their last collision. It seems reasonable that this distance should be related to the average distance a molecule undergoes between collisions. This distance is called the "mean free path."

The average spacing between molecules is the cube root of the inverse of the number density:

$$l_0 = \sqrt[3]{1/n} \tag{12.16}$$

However, we shall see that the average distance between collisions is larger than this. Imagine for the moment that our particles are solid spheres. A collision will occur when the center of one particle enters the so-called "sphere of influence," as illustrated in Fig. 12.3. Assuming the two particles are the same size, the sphere of influence has diameter equal to the particle diameter. The actual path followed by a particle is quite random, as shown in Fig. 12.4. For simplicity, however, let us straighten the path out (Fig. 12.5) and ask what frequency of collisions a particle undergoes when it is traveling at some constant speed, say the average speed. As the particle moves, its sphere of influence sweeps out a cylindrical volume per unit time of

$$\pi d^2 \bar{c} \tag{12.17}$$

with units of $[L^3/t]$. Given that the number density is n, the number of collisions per unit time experienced by the particle is

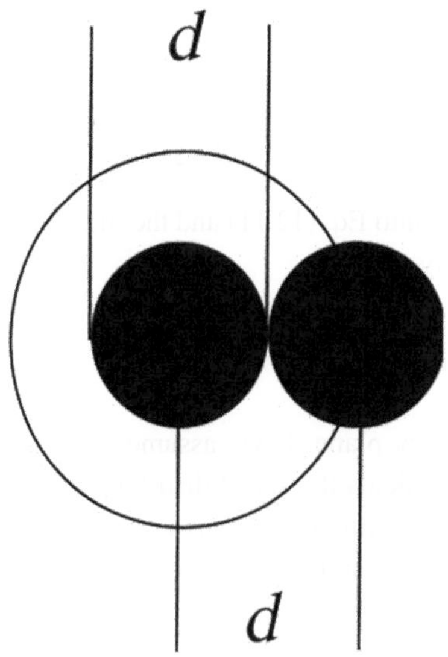

Figure 12.3 The sphere of influence.

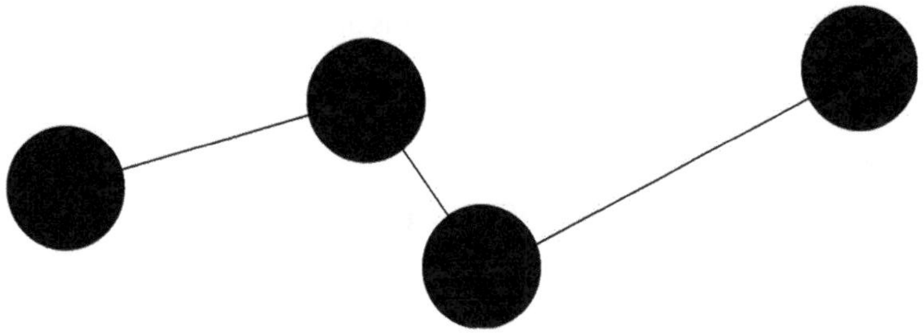

Figure 12.4 Random path of colliding particle.

Figure 12.5 Straightened path of colliding particle.

$$\nu_c = \pi d^2 \bar{c} n \tag{12.18}$$

This is called the "collision frequency" and has dimensions of $[1/t]$. Better theory for the collision frequency gives

$$\nu_c = \sqrt{2}\pi d^2 \bar{c} n \tag{12.19}$$

The average distance traveled by our particle, the mean free path, should be the average speed divided by the collision frequency, or

$$l = \frac{\bar{c}}{\nu_c} = \frac{1}{\sqrt{2}\pi d^2 n} \tag{12.20}$$

Note that the ratio of the mean free path to the average particle spacing is therefore

$$\frac{l}{l_0} = \frac{n^{1/3}}{\sqrt{2}\pi d^2 n} = \frac{1}{\sqrt{2}\pi d^2 n^{2/3}} \tag{12.21}$$

A good estimate of the size of an atom or molecule is its Lennard–Jones diameter.

12.1.2 Example 12.1

Consider diatomic nitrogen, N_2. For $p = 1$ bar and 298.15 K, calculate the number density, average molecular speed, collision frequency, average spacing, and mean free path.

The number density can be calculated using the ideal gas law:

$$n = \frac{kT}{p} = 2.46 \times 10^{25}\ \text{m}^{-3}$$

The average molecular speed is

$$\bar{c} = \sqrt{\frac{8kT}{\pi m}} = 473.2\ \text{m/sec}$$

From Table C.2, the collision diameter of N_2 is 3.74 Å. Thus, the collision frequency is

$$\nu_c = \sqrt{2}\pi n d^2 \bar{c} n = 7.24 \times 10^9 \text{ sec}^{-1}$$

The average spacing is

$$l_0 = \sqrt[3]{1/n} = 3.43 \times 10^{-9} \text{ m}$$

and the mean free path is

$$l = \frac{\bar{c}}{\nu_c} = \frac{1}{\sqrt{2}\pi d^2 n} = 6.53 \times 10^{-8} \text{ m}$$

Thus, their ratio is 19.0.

Returning to our transport problem, assume

$$\delta y = \pm \alpha l \tag{12.22}$$

where α is a constant of order unity. Therefore, the net flux of property ϕ becomes

$$J_\phi = \frac{n\bar{c}}{4}[\phi(y_0 - \alpha l) - \phi(y_0 + \alpha l)] \tag{12.23}$$

If the departure from equilibrium is not too severe, then the $\pm\delta y$ should be small compared to y_0. Therefore, let us expand ϕ in a Taylor series:

$$\phi(y_0 + \delta y) = \phi(y_0) + \frac{d\phi}{dy}\delta y + \cdots \tag{12.24}$$

and

$$\phi(y_0 - \delta y) = \phi(y_0) - \frac{d\phi}{dy}\delta y + \cdots \tag{12.25}$$

Keeping the first-order terms, J_ϕ becomes

$$J_\phi = -\frac{\alpha}{2} n\bar{c}l\frac{d\phi}{dy} \tag{12.26}$$

12.1.2.1 Momentum Transport

The average momentum carried per particle is

$$\phi = mu(y) \tag{12.27}$$

Its transport results in a shear stress, so that

$$J_{mom} = -\tau = -\frac{\alpha_\mu}{2} mn\bar{c}l\frac{du}{dy} = -\mu\frac{du}{dy} \tag{12.28}$$

One can show from more advanced theory that $\alpha_\mu \simeq 1$. Thus,

$$\mu = \frac{1}{2}\rho\bar{c}l \tag{12.29}$$

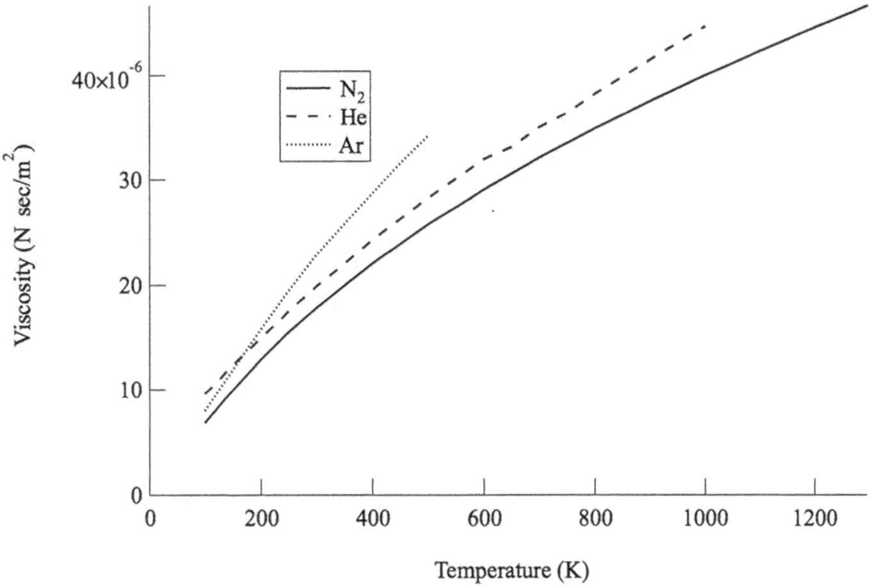

Figure 12.6 Viscosity of several gases.

Using the values from the numerical example above for N_2, we get 1.78×10^{-5} N sec/m^2, which is very close to the experimental value. Using the ideal gas law, one can show that

$$\mu \propto \frac{mT^{1/2}}{d^2} \tag{12.30}$$

This simple theory predicts that the viscosity scales as $T^{1/2}$. For N_2 this prediction is excellent, as shown in Fig. 12.6. However, it is not universally the case. The viscosity of helium, for example, scales more closely with $T^{0.65}$ or $T^{0.7}$, as also shown in the figure. In addition to temperature dependence, the simple theory predicts that viscosity is inversely proportional to collision cross-section. Since the ratio of the cross-section for He to N_2 is 0.465, and taking into account mass differences, one would expect the viscosity ratio to be about 0.81, whereas the actual ratio is 1.1, largely because collisional behavior is more complex than represented here. Thus, the simple theory is useful for a first-order understanding, but not for practical application.

12.1.2.2 Thermal Energy Transport

The average internal energy u carried per molecule for a monatomic gas is

$$\phi = u = mc_v T \tag{12.31}$$

Therefore, the heat flux is

$$J_q = q = -\frac{\alpha_\lambda}{2} mn\bar{c}lc_v \frac{dT}{dy} = -\lambda \frac{du}{dy} \tag{12.32}$$

and the thermal conductivity is thus

$$\lambda = \frac{\alpha_\lambda}{2}\rho\bar{c}lc_v = \frac{\alpha_\lambda}{\alpha_\mu}\mu c_v \tag{12.33}$$

Recall that the Prandtl number is

$$Pr = \frac{c_p\mu}{\lambda} \tag{12.34}$$

and that the ratio of specific heats is $\gamma = c_p/c_v$, so that

$$Pr = \frac{\alpha_\mu}{\alpha_\lambda}\gamma \tag{12.35}$$

For a monatomic gas $\gamma = 5/2$ and $Pr = 2/3$, therefore $\alpha_\mu/\alpha_\lambda = 5/2$. Thus

$$\lambda = \frac{5}{2}\mu c_v \tag{12.36}$$

The situation for non-monatomic gases is somewhat more complex. One simple approach was given by Eucken [45]. If we write the internal energy as

$$u = u_{tr} + u_{int} \tag{12.37}$$

then

$$c_v = c_{vtr} + c_{vint} \tag{12.38}$$

If we likewise assume that we can split the thermal conductivity, then we obtain the following expression:

$$\lambda = \lambda_{tr} + \lambda_{int} = \frac{5}{2}\mu c_{vtr} + \mu c_{vint} \tag{12.39}$$

While this expression illustrates the physics to some degree, it is not particularly accurate. For practical use it is better to use more advanced theory or experimental data.

12.1.2.3 Mass Transport

Mass transport is somewhat more complex. As we shall see below, when one has a multicomponent mixture in general the flux of any one species depends on the gradient of all the other species. However, if we consider the diffusion of a minor species with respect to the major species in a binary mixture, then Fick's Law applies and our simple approach can yield useful results. In this case, the sum of the number densities is

$$n_1 + n_2 = n \tag{12.40}$$

Therefore, assuming n is constant,

$$\frac{dn_1}{dy} = -\frac{dn_2}{dy} \text{ and } J_{n1} = -J_{n2} \tag{12.41}$$

Here, $\phi_1 = n_1/n$ is the fraction of species 1 transported by $n\bar{c}/4$. Thus,

$$J_{n1} = -\frac{1}{2}\alpha_n n\bar{c}l_1\frac{d\phi_1}{dy} = -\frac{1}{2}\alpha_n\bar{c}l_1\frac{dn_1}{dy} \tag{12.42}$$

Likewise,

$$J_{n2} = -\frac{1}{2}\alpha_n \bar{c} l_2 \frac{dn_2}{dy} \tag{12.43}$$

The diffusion coefficient is defined by the relation

$$J_{n1} = -D_{12}\frac{dn_2}{dy} \tag{12.44}$$

so that

$$D_{12} = \frac{1}{2}\alpha_n \bar{c} l_1 = D_{21} = \frac{1}{2}\alpha_n \bar{c} l_2 \tag{12.45}$$

The result that $D_{12} = D_{21}$ implies that $\sqrt{m}\pi d^2$ is the same for the two gases. This may not be true unless the two gases are nearly identical. As a result, our analysis is only valid for species 1 if $n_1 << n_2$.

12.2 The Boltzmann Equation and the Chapman–Enskog Solution

A full description of the flow field involves complete specification of the spatial distribution of the molecular velocities via the distribution function. Any property ϕ is a function of both velocity and space, known as "phase space":

$$\phi = \phi(c_i, x_i) \tag{12.46}$$

In general, the species-specific velocity distribution function f_k is described by Boltzmann's equation (Boltzmann's equation is derived in Appendix E and all terms defined and described):

$$\frac{\partial f_k}{\partial t} + c_j\frac{\partial f_k}{\partial x_j} + \frac{\partial}{\partial x_j}\left[F_{k_j} f_k\right] = \left.\frac{\partial f_k}{\partial t}\right)_{coll} \tag{12.47}$$

The terms in the equation account for time, transport in velocity space, body forces, and collisions. The collision term accounts for changes in molecular velocities due to collisions and, in reacting flow, due to collisions that result in reaction. For example, for binary collisions between monatomic species, the collision term takes on the form

$$\left.\frac{\partial f_k}{\partial t}\right)_{coll} = \sum_{j=1}^{N}\int\int\int (f_k' f_j' - f_k f_j) g_{jk} b\, db\, d\epsilon\, d\vec{c} \tag{12.48}$$

where the prime indicates quantities after the collision. Therefore, Boltzmann's equation is an integro-differential equation and is difficult to solve.

Chapman and Enskog independently solved Boltzmann's equation in the Navier–Stokes limit between 1910 and 1920. They proceeded by expanding the distribution function about the equilibrium values, obtaining equations for the expansion coefficients and then solving the resulting equations, also using expansion techniques. The details of the method can be found in Chapman and Cowling [46], Hirschfelder, Curtiss, and

Bird [14], Vincenti and Kruger [2], and Williams [47]. Their method resulted in expressions that depend on the exact form of the interaction potentials. The most common next step has been to use the Lennard–Jones potential to derive practical expressions for the transport coefficients. Here we merely quote the results.

12.2.1 Momentum Diffusion

From Chapman–Enskog theory the shear stress tensor is given by

$$\tau_{ij} = \mu \left(\frac{\partial u_i}{\partial x_j} + \frac{\partial u_j}{\partial x_i} \right) + \mu_B \delta_{ij} \frac{\partial u_l}{\partial x_l} \tag{12.49}$$

As before, μ is the viscosity and μ_B is the bulk viscosity. For ideal monatomic gases, μ_B is zero. However, it can be important in sound propagation and shock waves when internal energy is in non-equilibrium.

For parallel flow and ignoring bulk viscosity effects,

$$\tau_{xy} = -\mu \frac{du_x}{dy} \tag{12.50}$$

From Chapman–Enskog theory,

$$\mu = \frac{5}{16} \frac{\sqrt{\pi m k T}}{\pi d^2 \Omega_{ij}^{2,2}} \tag{12.51}$$

where $\Omega_{ij}^{(2,2)}$ is the collision integral for momentum transport (Appendix E.1).

For mixtures the picture is very complicated. See, for example, Eq. (8.2-25) in Hirschfelder, Curtiss, and Bird [14]. Therefore, it is usual in practical application to use an empirical mixing formula such as

$$\mu_{\text{mixture}} \simeq \sum_i^N \left\{ \frac{x_i}{\left[\frac{x_i}{\mu_i} + 1.385 \sum_{j \neq i}^N \left(\frac{x_j k T}{p m_i D_{ij}} \right) \right]} \right\} \tag{12.52}$$

12.2.2 Example 12.2

Calculate the viscosity for helium using the simple expression of Eq. (12.29) and the Chapman–Enskog expression of Eq. (12.51) from 300–1000 K and compare the results with the expression in Petersen [48] at 1 atm.

The appropriate expression from Petersen in the low-pressure limit is

$$\mu = 3.674 \times 10^{-7} T^{0.7}$$

This is a simple calculation, easily done in Matlab or Mathematica. One must calculate the collision integral. The simplest way is to interpolate from Table E.1. In the calculations used to generate the results shown in the figure, the author entered Table E.1 into a csv file, read the file in Matlab, and used the interp1 function. A plot of the result is shown below.

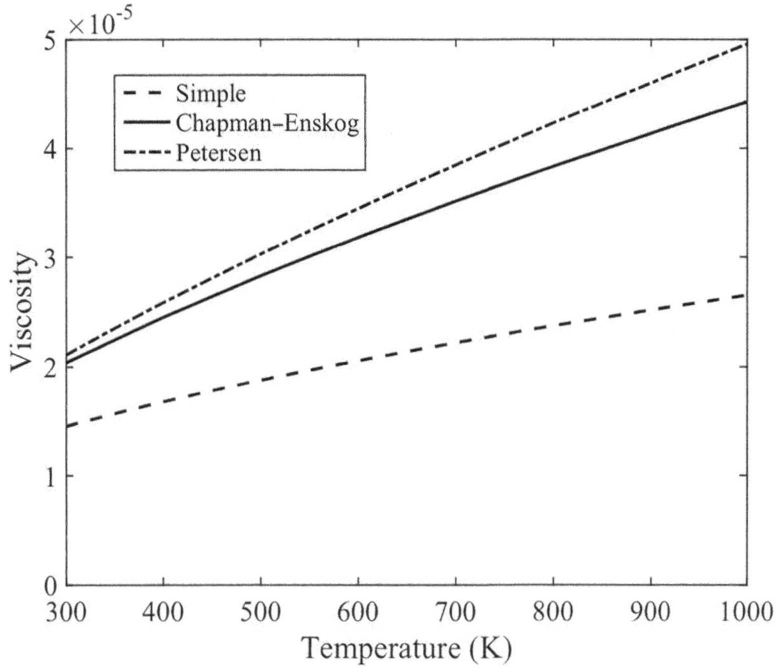

Figure 12.7 Viscosity of helium.

12.2.3 Thermal Diffusion

Kinetic theory shows that for a single-component monatomic gas

$$\vec{q} = \lambda \vec{\nabla} T \tag{12.53}$$

where the thermal conductivity λ is given by

$$\lambda = \frac{25}{16}\left(\frac{kTc_v}{\Omega_{ij}^{(2,2)}}\right) = \frac{5}{2}\mu c_v \tag{12.54}$$

($\Omega_{ij}^{(2,2)}$ is the collision integral for momentum transport, see Appendix E.1).

For mixtures the picture is again very complicated. [Again, see Eq. (8.2-25) in Hirschfelder, Curtiss, and Bird [14].] Chapman–Enskog gives

$$\vec{q}_{\text{mixture}} = -\lambda \vec{\nabla} T + \rho \sum_{i=1}^{N} h_i y_i \vec{v}_i + RT \sum_{i=1}^{N} \sum_{j=1}^{N} \left(\frac{x_j D_{T,i}}{W_i D_{ij}}\right)(\vec{v}_i - \vec{v}_j) + \vec{q}_R \tag{12.55}$$

where W is the molecular weight.

Therefore, as with viscosity, it is usual in practical application to use an empirical mixing formula such as

$$\lambda = \frac{1}{2}\left(\sum_{k=1}^{K} x_k \lambda_k + \frac{1}{\sum_{k=1}^{K} x_k/\lambda_k}\right) \tag{12.56}$$

12.2.4 Example 12.3

Calculate the thermal conductivity for helium using the simple expression of Eq. (12.36) and the Chapman–Enskog expression of Eq. (12.54) from 300–1000 K and compare the results with the expression in Petersen [48] at 1 atm.

The appropriate expression from Petersen in the low-pressure limit is

$$\lambda = 2.682 \times 10^{-3} T^{0.71}$$

This is a simple calculation, easily done in Matlab or Mathematica. A plot of the result is shown below.

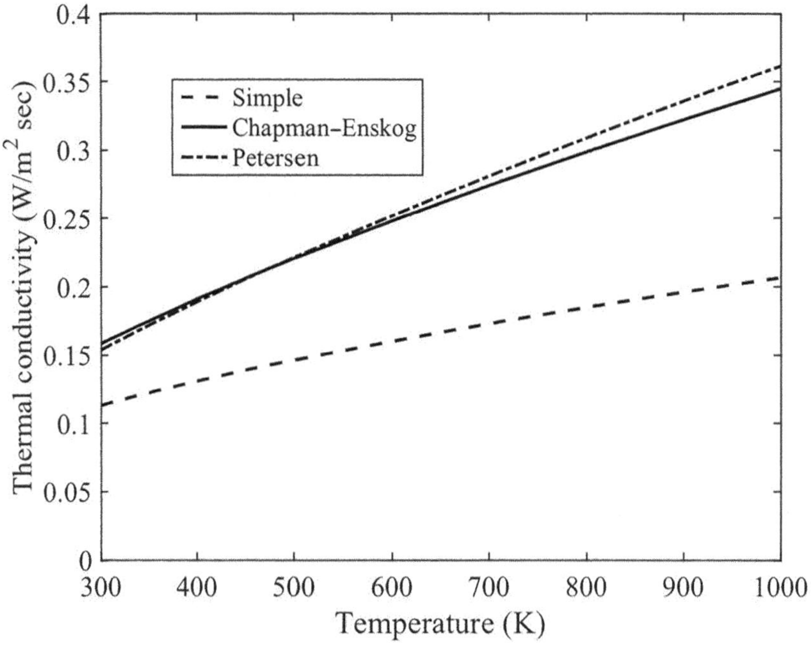

Figure 12.8 Thermal conductivity of helium.

12.2.5 Mass Diffusion

In the simple analysis of mass diffusion outlined in Section 12.1.2.3 the mass flux is simply proportional to the gradient of the species. In the more complete theory, things are considerably more complex. In general:

$$\vec{\nabla} x_i = \sum_{j=1}^{N} \left(\frac{x_i x_j}{D_{ij}} \right) (\vec{v}_j - \vec{v}_i) + (y_i - x_i) \frac{\vec{\nabla} p}{p} + \left(\frac{\rho}{p} \right) \sum_{j=1}^{N} y_i y_j (\vec{F}_i - \vec{F}_j)$$
$$+ \sum_{j=1}^{N} \left[\left(\frac{x_i x_j}{\rho D_{ij}} \right) \left(\frac{D_{T_j}}{y_j} - \frac{D_{T_i}}{y_i} \right) \right] \left(\frac{\vec{\nabla} T}{T} \right) \tag{12.57}$$

Here x_i and y_i are the mole and mass fractions of species i, respectively. The F_i are the species-specific body forces per unit volume. The D_{ij} are the pairwise multicomponent diffusion coefficients and D_{T_i}, D_{T_j} the thermal diffusion coefficients. D_{ij} is related to the two mole fractions, the pressure, and inversely proportional to the reduced mass and collision frequency:

$$D_{ij} = x_i x_j \frac{p}{\mu_{ij} \nu_{ij}} \tag{12.58}$$

where

$$\mu_{ij} = \frac{m_i m_j}{m_i + m_j} \qquad \text{reduced mass}$$
$$\nu_{ij} = n_i n_j d_{ij}^2 \left(\frac{8\pi kT}{\mu_{ij}} \right)^{1/2} \qquad \text{collision frequency}$$

For binary mixtures, Eq. (12.57) reduces to

$$y_i \vec{v}_i = -D_{12} \left[\vec{\nabla} y_1 - \left(\frac{y_1 y_2}{x_1 x_2} \right) (y_1 - x_1) \left(\frac{\vec{\nabla} p}{p} \right) - \frac{(y_1 y_2)^2}{(x_1 x_2)} \left(\frac{p}{\rho} \right) (\vec{F}_1 - \vec{F}_2) \right] \tag{12.59}$$

In either the general case or a binary mixture, these equations form a set of simultaneous equations for the diffusion velocities and must generally be solved numerically. However, in the case of a binary mixture, if $\vec{F}_1 = \vec{F}_2$ or both are zero, and $\vec{\nabla} p = 0$ or $y_1 = x_1$, then

$$y_i \vec{v}_1 \simeq -D_{12} \vec{\nabla} y_1 \tag{12.60}$$

This is known as Fick's law of diffusion. It is almost always used for simplicity. For a multi-species mixture, if all the above assumptions hold, then the Stefan–Boltzmann equation applies:

$$\vec{\nabla} x_i \simeq \sum_{i=1} n \frac{x_i x_j}{D_{ij}} (\vec{v}_j - \vec{v}_i) \tag{12.61}$$

The diffusion coefficients require information about collisions. The collision frequency can be approximated as in the simple analysis as

$$\nu_{ij} = n_i n_j \sigma_{ij} \overline{v_{ij}} = x_i x_j n \sigma_{ij} \overline{v_{ij}} \tag{12.62}$$

So that

$$D_{ij} = \frac{p}{\mu_{ij} n^2 \sigma_{ij} \overline{v}_{ij}} = \frac{kT}{n \mu_{ij} \sigma_{ij} \overline{v}_{ij}} \tag{12.63}$$

The more exact Chapman–Enskog expression is

$$D_{ij} = \frac{3}{16} \frac{\sqrt{\frac{2\pi k^3 T^3}{\mu_{ij}}}}{p\pi d_{ij}^2 \Omega_{ij}^{(1,1)}} \tag{12.64}$$

where $\Omega_{ij}^{(1,1)}$ is the $l, s = 1, 1$ collision integral (see Appendix E.1).

12.2.6 Example 12.4

Calculate the self-diffusion coefficient for helium using the simple expression of Eq. (12.45) and the Chapman–Enskog expression of Eq. (12.64) from 300–1000 K and 1 atm. Compare the calculated values with data from Kestin et al. [49].

This is a simple calculation, easily done in Matlab or Mathematica. A plot of the result is shown below. It illustrates the difficulty in determining the diffusion coefficients.

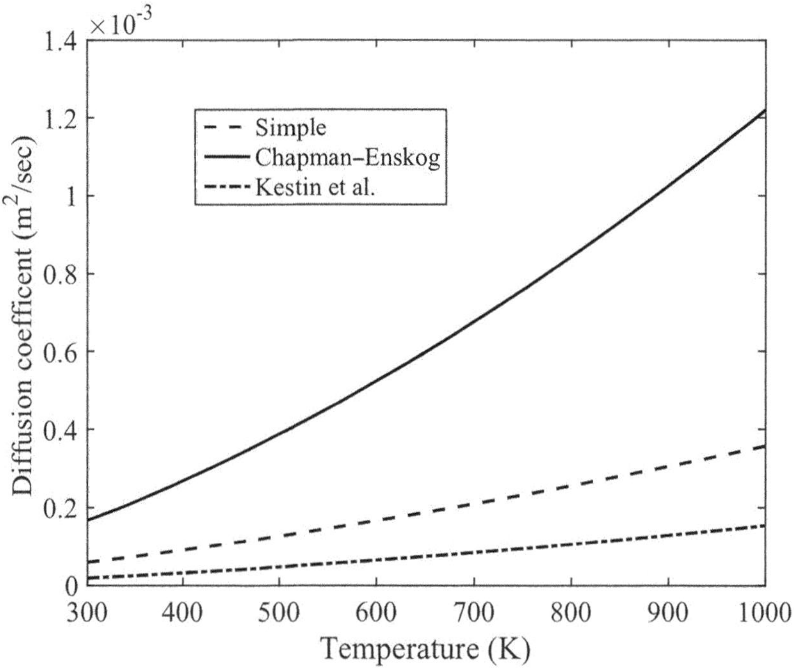

Figure 12.9 Self-diffusion coefficient of helium.

12.3 Transport Data Sources

From the examples above, which in each case compare the simple and Chapman–Enskog expressions with data, it should be clear that it is best to utilize sources for transport properties that have been validated using experimental data. Developing transport property data is part of NIST's mission, and one can obtain both thermodynamic and transport data for a large number of substances using Refprop, a NIST product, as well as the NIST Data Gateway. In addition, there are commercially available sources of transport data. The theories we have discussed in this chapter become valuable when quality data is not available, or to interpolate or extrapolate limited data sets.

12.4 Summary

This chapter is devoted to the topic of kinetic theory of gases. This is the study of the dynamic, non-equilibrium behavior of flowing systems that lead to the transport of momentum, energy, and mass.

Here we remembered the important transport properties: the shear stress tensor τ_{ij}, the heat flux vector q_i, and the species diffusion velocities v_{ki}. In the simple form you are probably most familiar with, these transport terms are typically written as

$$\tau = \mu \frac{du}{dy} \tag{12.65}$$

$$q = -\lambda \frac{dT}{dy} \tag{12.66}$$

$$J_1 = \rho_1 v_{1y} = -D_{12} \frac{dn_1}{dy} \tag{12.67}$$

where μ is the dynamic viscosity, λ the thermal conductivity, and D the binary diffusion.

12.4.1 Transport Phenomena

In this section we derived expressions for the transport properties using simple approximations. We introduced the concept of the mean free path

$$l = \frac{\bar{c}}{\nu_c} = \frac{1}{\sqrt{2}\pi d^2 n} \tag{12.68}$$

The important results that followed are for viscosity:

$$\mu = \frac{1}{2} \rho \bar{c} l \tag{12.69}$$

For a monatomic gas, the thermal conductivity:

$$\lambda = \frac{5}{2} \mu c_v \tag{12.70}$$

For mass diffusivity:

$$D_{12} = \frac{1}{2} \alpha_n \bar{c} l_1 = D_{21} = \frac{1}{2} \alpha_n \bar{c} l_2 \tag{12.71}$$

12.4.2 Boltzmann Equation and the Chapman–Enskog Solution

The expressions derived above are based on simplifying assumptions that are not very accurate. A more complete theory of transport is provided by the Chapman–Enskog solution to the Boltzmann equation. The expressions for momentum, energy, and mass transport coefficients are

$$\mu = \frac{5}{16} \frac{\sqrt{\pi mkT}}{\pi d^2 \Omega_{ij}^{2,2}} \tag{12.72}$$

$$\lambda = \frac{25}{16} \left(\frac{kTc_v}{\Omega_{ij}^{(2,2)}} \right) = \frac{5}{2} \mu c_v \tag{12.73}$$

$$D_{ij} = \frac{3}{16} \frac{\sqrt{\frac{2\pi k^3 T^3}{\mu_{ij}}}}{p\pi d_{ij}^2 \Omega_{ij}^{(1,1)}} \tag{12.74}$$

12.5 Problems

12.1 The Knudsen number plays an important role in low-density flow problems. It is defined as

$$K_n = \frac{\lambda}{L}$$

where λ is the mean free path and L a characteristic length. Flow for which $K_n \geq 1$ is called free molecular flow because very few collisions take place over the distance L. Consider a 30-cm sphere traveling through the atmosphere and take the diameter as the characteristic length. Find the altitude above which free molecular flow prevails. Assume a collision diameter of 3.7 Å and a temperature of 280 K.

12.2 The thickness of a laminar boundary layer at a given angular position on a circular cylinder of diameter D varies as $\delta/D \sim 1/\sqrt{Re}$ where Re is the Reynolds number, defined as $Re \equiv \rho UD/\mu$. What is the dependence of δ/D on temperature? Pressure?

12.3 An empirical correlation for the Nusselt number ($Nu = hd/k$) for flow past tubes is

$$Nu = 0.25 Re^{0.6} Pr^{0.38}$$

where $Re = \rho Ud/\mu$ is the Reynolds number and $Pr = c_p\mu/k$ is the Prandtl number. Here μ is the viscosity and k the thermal conductivity.

Using the simple expressions for the transport properties that we derived in class, determine the temperature and pressure dependency of the convective heat transfer coefficient h.

12.4 The simple form for viscosity can be written as

$$\mu = \frac{1}{2}\rho\bar{c}l = \frac{1}{\pi^{3/2}}\frac{\sqrt{mkT}}{d^2}$$

while the more complete Chapman–Enskog expression [Eq. (12.61)] is given by

$$\mu = \frac{5}{16\sqrt{\pi}}\frac{\sqrt{mkT}}{d^2\Omega_{ij}^{2,2}}$$

where d is the collision diameter and $\Omega_{ij}^{2,2}$ the collision integral given in Appendix E. Compare these two expressions numerically for argon and $T =$ 300–1000 K.

13 Spectroscopy

Spectroscopy is the practice of examining how light is absorbed, emitted, or scattered by atoms or molecules, and using the results to study atomic or molecular structure. If one examines absorption or emission across a range of wavelengths, the result is called a spectrum. The nature of the spectrum will be determined by the rate at which light is absorbed or emitted by individual atoms or molecules, and by the populations of the individual energy levels from which these interactions occur. For the engineer, spectroscopy is mostly of interest because of the ability to measure thermodynamic properties. Each of the types of spectroscopies we will discuss involves interrogating a medium with radiation and analyzing the radiation that is returned to a detector of some kind, or in the case of emission, observing light leaving the medium. For simplicity we will limit our discussion to making measurements of species in gases. Some excellent resources include Hanson, Spearrin, and Goldenstien [50] and Eckbreth [51].

As we learned from quantum mechanics, for atoms and molecules there exists a large number of energy levels or states. The spectral absorption and emission of radiation results from the transition of the atoms or molecules between states as the result of interacting with a radiation field. Scattering involves a dynamic process where initial and final states may be different. As we shall see, absorption and emission and fluorescence are resonant processes. Scattering (and this is largely a matter of definition) does not necessarily involve resonant interactions. Each of the methods we will discuss allow one to measure the populations of individual energy levels. Once the state populations are known, they can be directly related to temperature through the Boltzmann relation:

$$\frac{N_i}{N_j} = \frac{g_i}{g_j} \exp\left[(E_i - E_j)/kT\right] \tag{13.1}$$

Pressure can be measured in a variety of ways. If the temperature and density are known, then the pressure can be obtained from the perfect gas law. Under certain conditions the absorption line shape is a function of pressure. Using Rayleigh scattering, or a combination of Raman and Rayleigh scattering, it may be possible to measure the total number density, which combined with a temperature measurement and the perfect gas law allows determination of the pressure.

13.1 The Absorption and Emission of Radiation

A simple view of absorption or emission is that of interaction of atoms or molecules with photons, or quanta of radiant energy. In this view, the interaction is much like a collisional process, and the overall rate of the process is given by the product of a rate coefficient and the concentrations of the atoms/molecules and the photons. Illustrated in Fig. 13.1, this was first proposed by Einstein [52] in 1917. Radiative processes that involve causing a transition between two quantum states include absorption, induced emission, and spontaneous emission. Induced emission arises when the presence of a radiation field forces the molecule to a lower energy state. Spontaneous emission occurs as the result of the natural instability of excited states. Because these interactions involve photons whose energy is equal to the energy difference between specific energy levels, $E = h\nu$, they are called a "resonant" process.

It can be shown that the rate of absorption per molecule between states 1 and 2 is

$$W_{12} = \int_0^\infty B_{12} E_\nu(\nu)\phi(\nu)d\nu \tag{13.2}$$

where B_{12} is the Einstein coefficient for absorption, $E_\nu(\nu)$ the spectral irradiance, and $\phi(\nu)$ the normalized spectral line shape function. If $E_\nu(\nu)$ varies slowly with respect to the line shape function, then

$$W_{12} \cong B_{12} E_\nu(\nu) \tag{13.3}$$

As W_{12} contains the rate coefficient and, in essence, the photon concentration, the overall rate at which absorption takes place per unit volume is

$$W_{12} N_1 \tag{13.4}$$

where N_1 is the number density, or population of state 1. Likewise, the rates per molecule of induced emission and spontaneous emission are

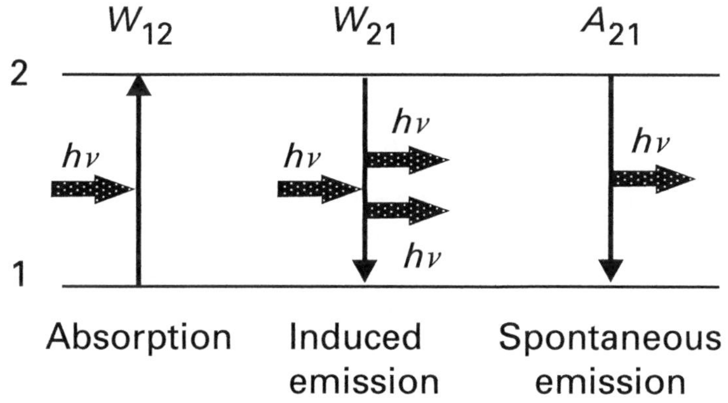

Figure 13.1 The Einstein radiative processes.

$$W_{21} = \int_0^\infty B_{21} E_\nu(\nu)\phi(\nu)d\nu \tag{13.5}$$

and

$$A_{21} \tag{13.6}$$

where B_{21} is the Einstein coefficient for induced emission and A_{21} the Einstein coefficient for spontaneous emission. Thus, the overall rates are

$$W_{21}N_2 \text{ and } A_{21}N_2 \tag{13.7}$$

Using equilibrium thermodynamic arguments and the principle of detailed balancing, Einstein showed that for transitions between two states i and j, there is a fixed relationship between all three coefficients

$$\frac{A_{ij}}{B_{ij}} = 8\pi h\nu^3/c^2 \tag{13.8}$$

and

$$g_i B_{ij} = g_j B_{ji} \tag{13.9}$$

where c is the speed of light and the g's are the degeneracies of the two states. Therefore, we can also write

$$g_1 W_{12} = g_2 W_{21} \tag{13.10}$$

The Einstein coefficients can be calculated from the relation

$$B_{ij} = \frac{8\pi^3 R^2}{3ch^2 g_i} \tag{13.11}$$

R^2 is the probability for the transition from state i to state j and is given by the square of the electric dipole moment μ:

$$R^2 = \left| \int \psi_i \vec{\mu} \psi_j dV \right|^2 \tag{13.12}$$

where

$$\vec{\mu} = \sum e\vec{r} \tag{13.13}$$

is the electric dipole moment operator, e is the charge of each particle, r is the distance between charges, and dV indicates integration over all space. If R has a reasonable finite value, then the radiative transition is said to be allowed. Otherwise, it is forbidden. An examination of R's leads to selection rules which determine what types of transitions are allowed.

Transition probability data is sometimes expressed in terms of the oscillator strength, f_{ij}. This is a non-dimensional number and is the ratio of the actual transition probability to the transition probability predicted by the classical harmonic oscillator model.

Einstein coefficients, transition probabilities, or oscillator strengths can be found tabulated in a number of sources. A large number of atomic transitions are listed in Sansonettia and Martin [53], who give both structure and spectroscopic data. Radzig and Smirov [9] give data for both atomic and molecular transitions in terms of excited state lifetimes, which is the inverse of the sum of the Einstein A coefficients for allowed transitions from the given state. The table below lists typical values of excited state lifetimes for some important atomic and molecular electronic transitions.

Atom or molecule	Ground state	Excited state	Lifetime (ns)
Na	$3s(^2S_{1/2})$	$3p\left(^2P^0_{1/2}\right)$	16.4
O	$2p^4(^3P_2)$	$3s'\left(^1D^0_2\right)$	2.0
OH	$X^2\Pi_i$	$A^2\Sigma^+$	0.00069
CH	$X^2\Pi_r$	$A^2\Delta$	0.0005
C_2	$X^1\Sigma_g^+$	$C^1\Pi_g$	0.000031
NO	$X^2\Pi_r$	$B^2\Pi_r$	0.0031

For molecules, the individual Einstein coefficients are a function not only of the electronic states involved, but of the vibrational and rotational states as well. Therefore, to make accurate predictions of spectra, or conversely to properly interpret experimental spectra, one must use accurate transition probability data. LIFBASE [30], a Stanford Research Institute program, contains a compilation of structural and transition probability data, including Einstein A coefficients, for OH, OD, NO, CH, and CN.

13.2 Spectral Line Broadening

When the spectral nature of light that has been absorbed or emitted by a particular atom or molecule is examined in detail, it is found that the light is absorbed or emitted over a finite bandwidth, contrary to what would be expected from a simple quantum-mechanical point of view. This effect is called line broadening.

Line broadening in a gas medium arises for three main reasons. The first is due to the finite lifetime of individual atoms or molecules when they are in excited quantum states. This is called lifetime or natural broadening. The second is because of collisions between particles which limit the amount of time available for the particle to radiate without disturbance. This is called pressure broadening, because the collision frequency an individual particle experiences is proportional to the pressure. The third reason for broadening in flames is the random motion of individual atoms or molecules. This random motion, coupled with the Doppler effect, leads to Doppler broadening. It is conventional to introduce the spectral distribution due to line broadening as a normalized line shape function, $\phi(\nu)$. By normalized we mean that the integral of the line shape function over frequency space is unity. For many purposes, line broadening can be described by two distribution functions.

The Fourier transform of the time-dependent power radiated by an oscillating atom or molecule is a delta function if the molecule is allowed to radiate for an infinite time. If the oscillation is truncated, however, one may show that the Fourier transform of the power signal leads to a frequency distribution that is a Lorentzian function of the form

$$\phi(\nu) = \frac{1}{\pi} \frac{\Delta\nu_L/2}{(\nu - \nu_0)^2 + (\Delta\nu_L/2)^2} \tag{13.14}$$

where $\Delta\nu_L = 1/\tau$ is the Lorentzian full width at half maximum (FWHM) and τ is the period that the particle has been allowed to radiate, about equivalent to the radiative lifetime or inverse of the collision frequency. (In practice, the line widths are determined experimentally.)

Doppler broadening is described by a Gaussian function:

$$\phi(\nu) = \left(\frac{mc^2}{2\pi kT\nu_0^2}\right) \exp\left[\frac{-mc^2(\nu - \nu_0)}{2kT\nu_0^2}\right] \tag{13.15}$$

One can show that

$$\Delta\nu_D = \sqrt{\frac{8kT\nu_0^2 \ln(2)}{mc^2}} \tag{13.16}$$

is the Doppler broadened FWHM.

If Doppler and pressure broadening are comparable in width, then a convolution of the two, called the Voigt profile, is used:

$$\phi(\nu)_V = \frac{a}{\pi\sqrt{\pi}} \frac{2\sqrt{\ln 2}}{\Delta\nu_D} \int_{-\infty}^{\infty} \frac{e^{-y^2}}{(\xi - y)^2 + a^2} dy \tag{13.17}$$

where

$$a = \frac{\Delta\nu_L}{\Delta\nu_D}\sqrt{\ln 2} \tag{13.18}$$

$$\xi = \frac{2(\nu - \nu_0)\sqrt{\ln 2}}{\Delta\nu_D} \tag{13.19}$$

$$y = \frac{2(\nu' - \nu_0)\sqrt{\ln 2}}{\Delta\nu_D} \tag{13.20}$$

Here ν' is the dimensional dummy variable in the integration, and ν_0 the unshifted line center. It is conventional to write the Voigt profile in the form

$$\phi(\nu) = \frac{2\sqrt{\ln 2}}{\sqrt{\pi}\Delta\nu_D} V(a, \xi) \tag{13.21}$$

where

$$V(a, \xi) = \frac{a}{\pi} \int_{-\infty}^{\infty} \frac{e^{-y^2}}{(\xi - y)^2 + a^2} dy \tag{13.22}$$

is called the Voigt function.

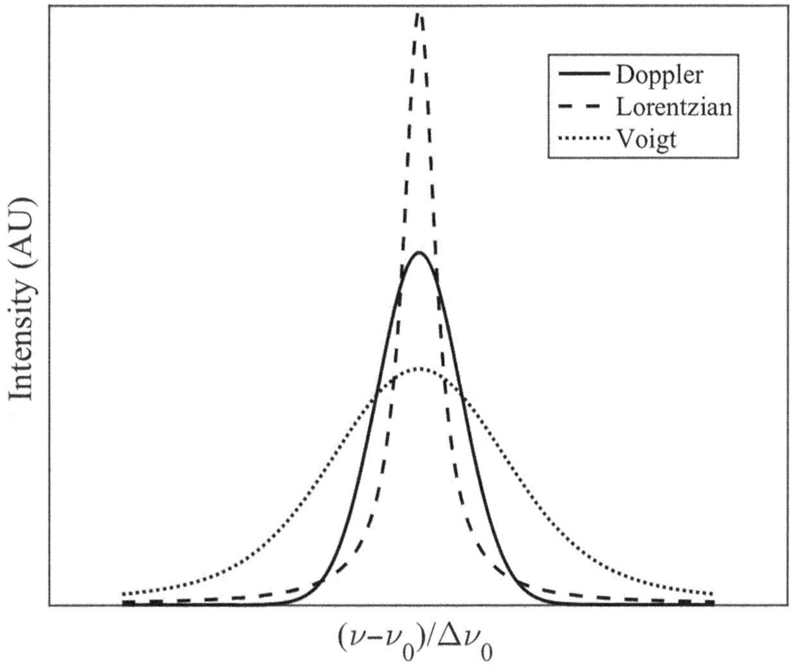

Figure 13.2 Spectral line shape functions.

All three distributions are illustrated in Fig. 13.2 for the case where the Doppler and Lorentz widths are equal. Note that the Lorentzian distribution is much broader than the Doppler distribution. Thus, when $\Delta\nu_L$ and $\Delta\nu_D$ are of similar magnitude, the Lorentzian will dominate in the wings and the Doppler at line center. The Voigt function is tabulated in a number of sources, including Mitchell and Zemansky [54]. For computational purposes, Humlicek [55] has presented an efficient algorithm for evaluating the Voigt function. For high pressures the situation is more complex.

13.3 Atomic Transitions

The only possible transitions for atoms involve changes in electronic quantum state. In practice, this means that most transitions occur in the near-infrared, visible, and near-ultraviolet regions of the electromagnetic spectrum between about 1000 Å and 1 μm. In atoms there is no restriction on the change in principal quantum number. However, the orbital angular momentum change must correspond to

$$\Delta L = 0 \text{ or } \pm 1 \tag{13.23}$$

Furthermore, when the transition arises from a change in state of a single optical electron, its orbital angular momentum can only change $\Delta l = \pm 1$. The total angular momentum quantum number can only change $\Delta J = 0$ or $\pm\ 1$. Transitions between two states with $J = 0$ are forbidden. The selection rule for S is $\Delta S = 0$. That is, S cannot change.

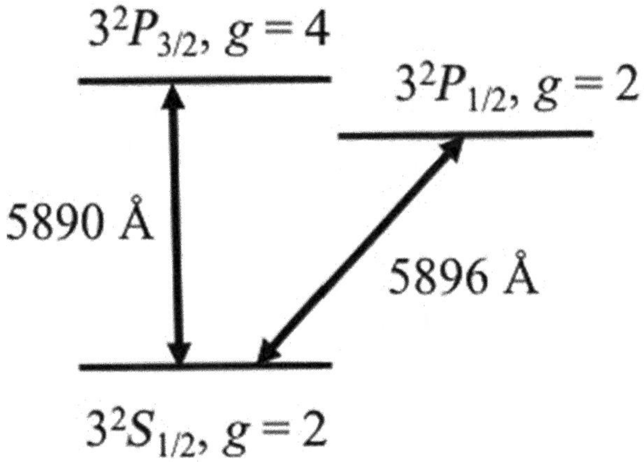

Figure 13.3 Sodium energy-level diagram.

For larger atoms the picture is more complicated and these rules begin to break down. Figure 13.3 shows the energy-level diagram for the lower energy levels of sodium. Radiative transitions are shown as arrows between energy levels. The number on each arrow is the wavelength of the transition in angstroms. The two transitions at 5890 and 5896 Å are the famous sodium D-lines. Their wavelength places them in the orange portion of the visible spectrum. Sodium lamps are sometimes used for street lights and will have an orange tinge.

13.4 Molecular Transitions

The spacings of electronic, vibrational, and rotational energy levels in molecules are separated by factors roughly of 100. Thus, molecular spectra appear in three distinct regions of the electromagnetic spectrum. Pure rotational transitions between states in the same vibrational and electronic levels appear in the microwave region, about 0.1 mm to 30 cm in wavelength. Vibration–rotation transitions, in which the vibrational and rotational states can change but not the electronic state, appear in the infrared region, between about 1 and 100 μm.

Selection rules for molecular transitions depend on the details of the particular molecule and whether simplifications like the rigid rotator or harmonic oscillator can be made. Because of the Born–Oppenheimer approximation, the wave function can be separated into electronic, vibrational, and rotational components:

$$\psi = \psi_e \psi_v \psi_r \tag{13.24}$$

As a result, the transition probability given above can be approximated as

$$R^2 = \left| \int \psi_e' \psi_v' \vec{\mu} \psi_e'' \psi_v'' dV \int \psi_r' \psi_r'' dV \right| \tag{13.25}$$

where it is typical to denote the higher energy state with a single prime, and the lower with a double prime.

13.4.1 Rotational Transitions

For pure rotational transitions a spectrum will appear only if the permanent electric dipole moment is nonzero. When the electronic and vibrational states do not change, $\psi_e{}'\psi_v{}' = \psi_e{}''\psi_v{}''$, so that

$$R^2 = \left| \int R_e(r)\psi_v{}^2(r)dr \int \psi_r{}'\psi_r{}''dV \right| \tag{13.26}$$

The permanent electric dipole moment is defined as

$$R_e(r) = \int \mu\psi_e{}^2 dV \tag{13.27}$$

The permanent electric dipole moment is zero for homonuclear diatomic molecules like H_2 and O_2, and nonzero for heteronuclear diatomic molecules.

The selection rules are determined by the rotational overlap integral

$$\int \psi_r{}'\psi_r{}''dV \tag{13.28}$$

This integral is zero except for transitions where $\Delta J = \pm 1$. Since (for a rigid rotator)

$$G(J) = B_e J(J+1) \tag{13.29}$$

the absorption lines appear at wavenumbers of

$$\sigma(J', J'') = B_e \left\{ J'(J'+1) - J''(J''+1) \right\} \tag{13.30}$$

Figure 13.4 illustrates a rotational absorption band.

13.4.2 Vibrational Transitions

For vibrational transitions, R^2 becomes

$$R^2 = \left| \int \psi_v{}'R_e(r)\psi_v{}''dr \int \psi_r{}'\psi_r{}''dV \right| \tag{13.31}$$

Because the wave functions are orthogonal, the first integral disappears unless R_e is a function of r. If we expand R_e in a power series so that

$$R_e(r) \cong R_e(r_0) + \frac{\partial R}{\partial r}(r - r_0) + \cdots \tag{13.32}$$

then

$$R^2 \cong \left| \frac{\partial R}{\partial r} \int \psi_v{}'(r - r_e)\psi_v{}''dr \int \psi_r{}'\psi_r{}''dV \right|^2 \tag{13.33}$$

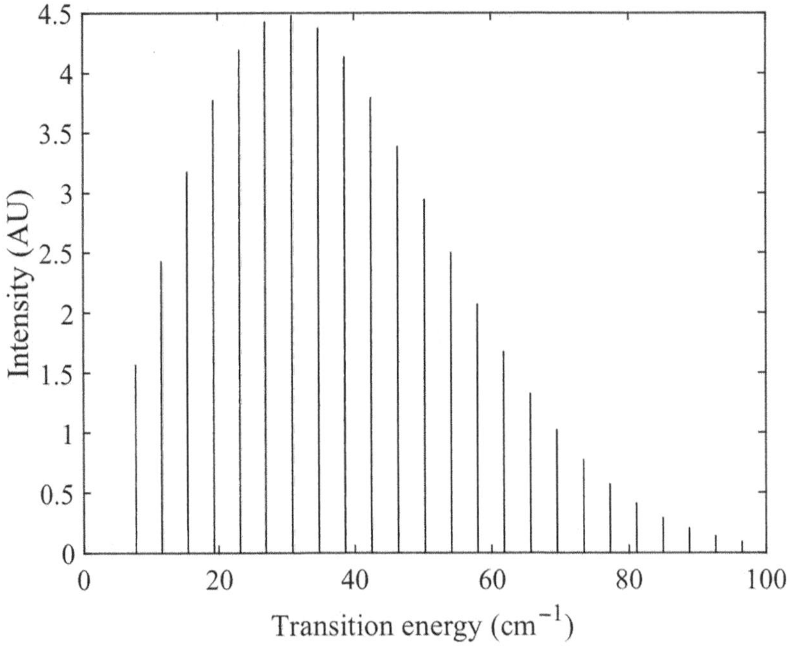

Figure 13.4 Rotational absorption spectrum.

If the wave functions are harmonic, the first integral in this expression is zero, except for $\Delta v = \pm 1$. Thus, the wavenumbers of the lines are

$$\sigma = \sigma_0 + B'J'(J'+1) - B''J''(J''+1) \tag{13.34}$$

where $\sigma_0 = G(v') - G(v'')$ is called the band origin. The rotational overlap integral still allows transitions for which $\Delta J = \pm 1$. The two possibilities for change in J are called the R and P branches, as $J' = J''+1$ or $J' = J''-1$, respectively. A vibrational–rotational absorption spectrum for which $B' = B''$ is shown in Fig. 13.5. Additional branches can arise depending on the complexity of the structure.

13.4.3 Electronic Transitions

For electronic transitions, the transition probability can be written as

$$R^2 = \left| \int \psi_v{}' R_e(r) \psi_v{}'' dr \int \psi_r{}' \psi_r{}'' dV \right| \tag{13.35}$$

where

$$R_e(r) = \int \psi_e{}' \mu \psi_e{}'' dV \tag{13.36}$$

is the electronic transition moment evaluated between the initial and final electronic states. Although it is a function of inter-nuclear distance, it does not usually change

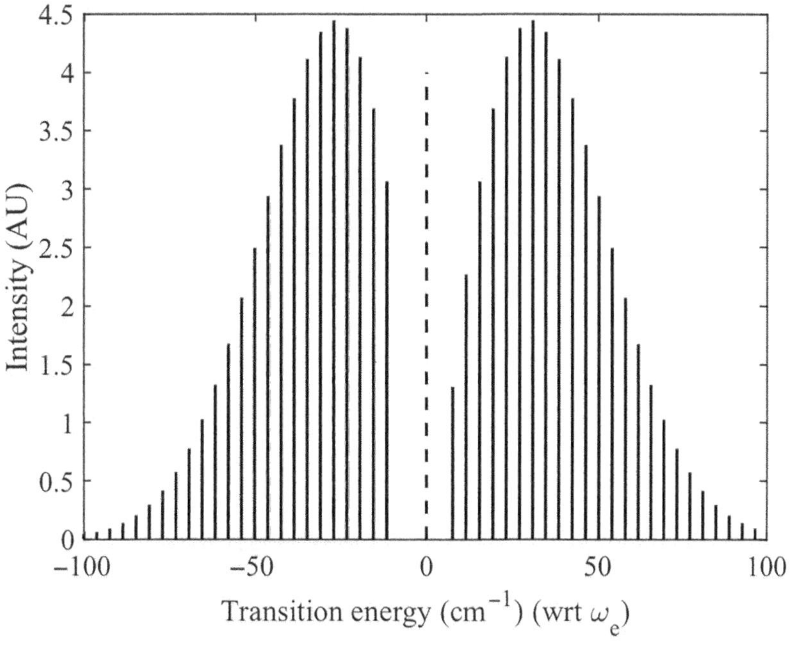

Figure 13.5 Vibrational/rotational absorption spectrum (horizontal axis is wavelength) (note that the J' to J'' transition that is shown dashed is forbidden; also, the rotational spacings are greatly exaggerated).

very rapidly with r. As a result, one can often approximate the transition probability by pulling R_e out of the vibrational overlap integral so that

$$R^2 = R_e^2(\bar{r}_{v'v''})q_{v'v''}S_{J'J''} \tag{13.37}$$

Here, $q_{v'v''}$ is called the Franck–Condon factor and $S_{J'J''}$ the line strength factor. These are defined as

$$q_{v'v''} = \left| \int \psi_v{}'\psi_v{}''dV \right|^2 \tag{13.38}$$

$$S_{J'J''} = \left| \int \psi_r{}'\psi_r{}''dV \right|^2 \tag{13.39}$$

$\bar{r}_{v'v''}$ is the r-centroid for the transition, and is defined as

$$\bar{r}_{v'v''} = \int \psi_v{}'r\psi_v{}''dr \tag{13.40}$$

The electronic spectrum is composed of bands that arise from transitions between specific vibrational and electronic states. Each pair of vibrational states may give rise to a band, which is composed of transitions between different rotational levels. The selection rules for electronic and vibrational transitions depend in detail on the above integrals. In general, the electronic selection rules

Table 13.1 Main transition bands

Title	Label	Transition	$S_{J'J''}$ (Emission)
R-branch	$R(J'')$	$J' = J'' + 1$	$\frac{(J'+\Lambda)(J'-\Lambda)}{J'}$
Q-branch	$Q(J'')$	$J' = J''$	$\frac{\Lambda^2(2J'+1)}{J'(J'+1)}$
P-branch	$P(J'')$	$J' = J'' - 1$	$\frac{(J'+\Lambda+1)(J'-\Lambda)+1}{J'+1}$

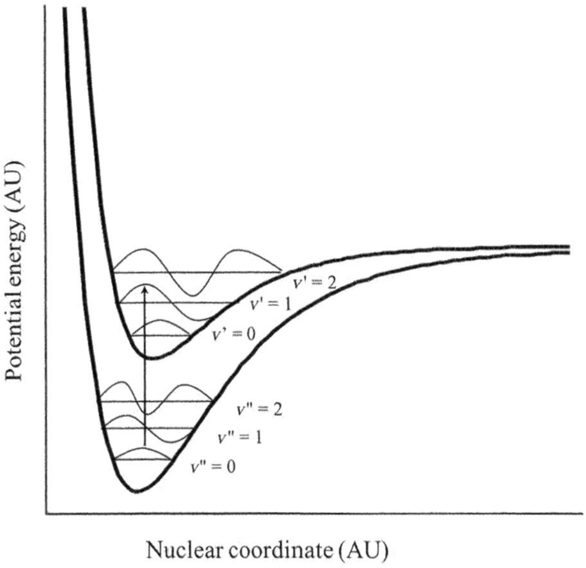

Figure 13.6 Illustration of the Franck–Condon principle.

$$\Delta\Lambda = 0, \ \pm 1 \tag{13.41}$$

and

$$\Delta S = 0 \tag{13.42}$$

hold. Depending on spin–orbit interactions, other rules may apply. In addition, numerous vibrational bands are allowed and are usually labeled as (v', v''). The Franck–Condon factor will be a maximum for those cases where maxima in the two wave functions overlap each other, as illustrated in Fig. 13.6.

For rotational energy changes the rule $\Delta J = \pm 1$ holds. However, if the orbital angular momentum of either electronic state is nonzero, then $\Delta J = \pm 0$ is also allowed. These transitions are usually labeled, as shown in Table 13.1. Also shown are the line strength factors for transitions between two electronic states with the same Λ and $S = 0$. Figure 13.7 illustrates the types of rotational/vibrational transitions allowed for a molecule with R, Q, and P branches. As molecular complexity grows, additional bands can appear.

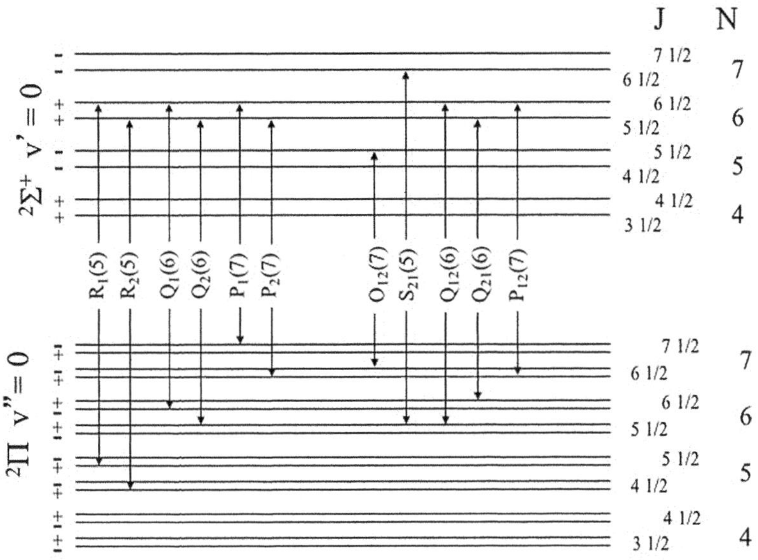

Figure 13.7 Electronic transitions resulting in a band spectrum. In this case, only $\Delta v = 0$ is allowed. The band locations and shapes depend on the detailed energy-level parameters. The $\Delta v = 0$ bands will overlap if the higher-order terms in $G(v)$ are zero. Note the presence of an additional band labeled "O" as a result of the more complicated structure of OH than accounted for by the simple theory. Also, for a molecule like OH, additional quantum numbers appear.

The line strength factors depend in detail on the electronic structure and additional transitions are possible depending on the structure. However, for a set of transitions from an excited state, the sum of line strength factors must be equal to the rotational degeneracy of the excited state:

$$\sum_{J''} S_{J'J''} = 2J'' + 1 \tag{13.43}$$

13.5 Absorption and Emission Spectroscopy

For absorption, light from a source is passed through the medium and modified by absorption due to individual atomic or molecular transitions, the intensity being diminished when individual absorption transitions take place. The character of the spectra will depend on spectral resolution and the types of transitions involved. Emission spectra are characterized by increases in intensity corresponding to individual transitions. Atomic spectra can be quite sparse, as atomic electronic transitions are usually widely spaced in energy. For molecules, the spectra are more congested because of rotational and vibrational spectra. In most cases, electronic transitions in both atoms and molecules are located in the visible ultraviolet region, molecular vibrational transitions in the infrared region, and rotational transitions in the microwave region.

To interpret the observed absorption and emission spectra requires solution of the radiative transport equation, which describes the passage of light through an absorbing and emitting medium. Consider the case of a single transition between states 1 and 2. Then, the transport of spectral irradiance in the z direction is given by

$$\frac{dI_\nu}{dz} = -h\nu[B_{12}N_1 - B_{21}N_2]\phi(\nu)I_\nu + \frac{A_{21}}{4\pi}N_2\phi(\nu) \tag{13.44}$$

The first term represents absorption, the second induced emission, and the third spontaneous emission. $\phi(\nu)$ is the normalized line shape function. It is conventional to define

$$\alpha_\nu \equiv h\nu[B_{12}N_1 - B_{21}N_2]\phi(\nu) \tag{13.45}$$

as the absorption coefficient, and

$$\epsilon_\nu \equiv h\nu\frac{A_{21}}{4\pi}N_2\phi(\nu) \tag{13.46}$$

as the emission coefficient. The radiative transfer equation becomes

$$\frac{dI_\nu}{dz} = -\alpha_\nu I_\nu + \epsilon_\nu \tag{13.47}$$

For the practical case of multiple transitions contributing to the spectrum, the absorption and emission coefficients become

$$\alpha_\nu = \sum_i \alpha_{\nu_i} \tag{13.48}$$

$$\epsilon_\nu = \sum_i \epsilon_{\nu_i} \tag{13.49}$$

where the summation is over all allowed transitions that absorb or emit in the bandwidth of interest.

The general solution to the radiative transport equation is

$$I_\nu(z) = e^{-\int_0^z \alpha_\nu dz}\left\{\int_0^z \epsilon_\nu e^{\int_0^z \alpha_\nu dz} dz + I_\nu(0)\right\} \tag{13.50}$$

The integral in the exponential is a very important quantity called the optical depth:

$$\tau_\nu = \int_0^z \alpha_\nu dz \tag{13.51}$$

For small τ_ν, a ray of radiation will be transmitted unaffected through the medium, and any radiation emitted within the medium will be unattenuated as it traverses to the boundary of the medium. On the other hand, if τ_ν is large, then no radiation entering the medium will pass through. These limits are called optically thin and optically thick.

If emission is not important, and assuming uniform properties through the medium, then

$$I_\nu(z) = I_\nu(0)e^{-\int_0^z \alpha_\nu dz} \tag{13.52}$$

Also, if the integral term is small enough, one can expand the exponential and retaining the first term, one obtains

$$I_\nu(z) = (1 - \alpha_\nu z)I_\nu(0) \tag{13.53}$$

This is known as Beer's Law. Therefore, assuming Beer's Law applies, and given the line shape function $\phi(\nu)$, one can recover the concentrations of the energy levels involved.

If emission is to be measured, and if absorption can be neglected, then

$$I_\nu(z) = \int_0^z \epsilon_\nu dz \tag{13.54}$$

Note that in real applications of absorption spectroscopy, integration along a line of sight is involved. If there is variation along the line of sight, only the average value of absorption is available, hence there is no ability to determine the spatial distribution of concentration, for example. In some cases, such as axial symmetry, it is less of a problem. Also, if complementary measurements reveal profile shapes, then corrections are possible. If the optical depth is too large, the incident light will be too severely attenuated to contain much useful spectral information.

In practical infrared spectroscopy, assuming a uniform medium, Eq. (13.52) can be written as

$$I_t(\nu) = I_0(\nu)e^{-k_\nu x} \tag{13.55}$$

where I_t and I_0 are the transmitted and incident light intensities, k_ν the spectral absorption coefficient, $x = pl$ where p is the partial pressure of the absorbing species and l the path length. The optical depth is then

$$\tau = k_\nu x \tag{13.56}$$

It was because of the line of sight problem that considerable effort has been expended over recent decades to develop measurement systems based on light scattering. Laser-induced fluorescence, Rayleigh and Raman scattering, and coherent anti-Stokes Raman scattering are examples.

13.5.1 Example 13.1

Consider measuring the concentration of CO in the atmosphere using absorption of a line in the ro-vibrational spectrum at 4.6 μm. The spectral absorption coefficient is 9.76 $\text{atm}^{-1}\ \text{cm}^{-1}$. The National Air Quality Standard for CO in the atmosphere is 9.0 ppm (parts per million). If an infrared laser is used as the light source, at what path length will 50% of the incident light be absorbed.

Solving Eq. (13.55) for x, one obtains

$$x = \frac{-\ln\left(I_t/I_0\right)}{k_\nu p_{CO}} = 7.1 \text{ km}$$

13.6 Laser-Induced Fluorescence

One method for overcoming the line of sight problem is laser-induced fluorescence (LIF). The method, illustrated in Figs 13.8 and 13.9, involves passing a beam of light, tuned in frequency to a specific absorption energy, through the medium. Absorption

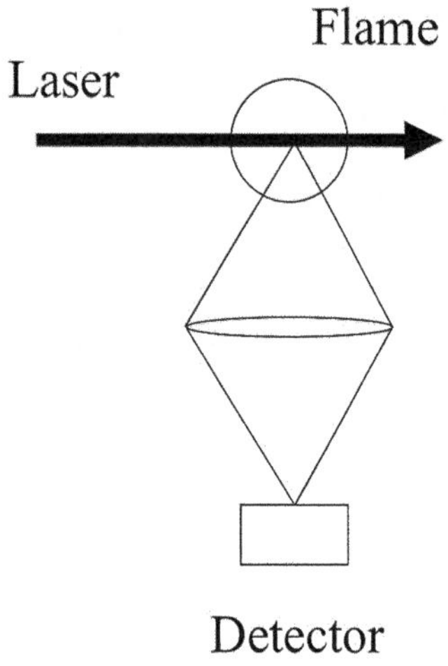

Figure 13.8 LIF optical arrangement.

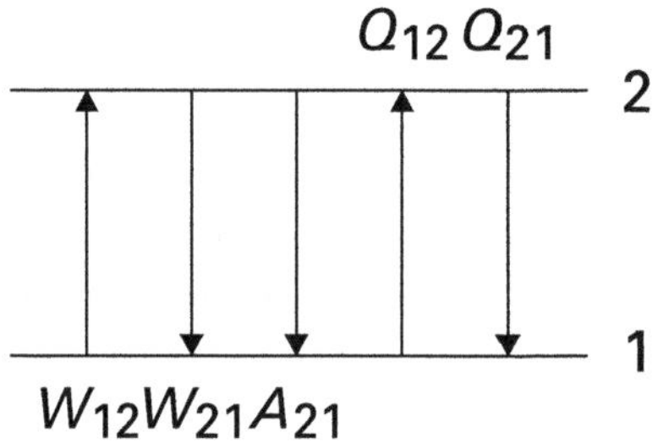

Figure 13.9 Two-level system showing radiative and collisional rate terms.

involves excitation of the atom or molecule to an excited state from which spontaneous emission can occur. The spontaneous emission will take place in all directions, including at right angles to the incoming beam. By imaging a portion of the laser beam, one samples the spontaneous emission from a volume defined by the diameter of the laser beam over a length determined by the collection optics.

The signal one measures is

$$Signal = C \int_{\Delta t} \int_{V_c} N_2(t) dV_c dt \tag{13.57}$$

where C is a calibration constant and V_c is the effective interaction volume defined by the optics. The integral is over the typical laser pulse width.

To relate the LIF signal to energy-level populations requires analyzing rate equations that describe the time-dependent populations of the energy levels involved during the laser pulse. In the two-level example, one must account for both the radiative terms and collisions. If Q_{12} and Q_{21} are the rates of collisional excitation and de-excitation, respectively, then the rate equations that describe the populations of the two states can be written as

$$\frac{dN_1}{dt} = (W_{21} + A_{21} + Q_{21})N_2 - (W_{12} + Q_{12})N_1 \tag{13.58}$$

$$\frac{dN_2}{dt} = (W_{12} + Q_{12})N_1 - (W_{21} + A_{21} + Q_{21})N_2 \tag{13.59}$$

In the absence of radiation, the ratio of the two populations is

$$\frac{N_2}{N_1} = \frac{Q_{12}}{Q_{21}} \tag{13.60}$$

However, this must be equated to Boltzmann's ratio, so that Q_{12} and Q_{21} are related by a detailed balance.

Noting that the total population is $N_T = N_1 + N_2$, the steady-state solution of Eqs (13.58) and (13.59) is

$$N_2 = \frac{Q_{12} + W_{12}}{Q_{12} + Q_{21} + A_{21} + W_{12} + W_{21}} N_T \tag{13.61}$$

If the laser intensity is adjusted so that $W_{12} >> Q_{12}$ but $(W_{12} + W_{21}) << (Q_{12} + Q_{21} + A_{21})$, then the population of state 2 becomes a linear function of the total population:

$$N_2 \sim W_{12} N_T \tag{13.62}$$

and thus the measured signal is proportional to the total population. This is called the linear limit and is how LIF is most typically used.

13.7 Rayleigh and Raman Scattering

Scattering usually refers to light/matter interactions where the photon energy is not resonant, or equal to, exact differences between quantum energy states. The excitation is typically provided by a laser source. Elastic scattering is the term used to describe a scattering process where the initial and final quantum states of the atom or molecule involved are the same. Rayleigh scattering, which causes the sky to look blue during the day, and red in early morning or late evening, is elastic scattering. Inelastic scattering involves processes that do result in a change of quantum state. Raman scattering is the most common such process. For both normal absorption and scattering, the process can be linear or nonlinear. Linear processes are ones where the signal one seeks is linearly proportional to both the light source intensity and the molecular concentration. Rayleigh and Raman scattering are linear processes. Examples of nonlinear processes are coherent Raman scattering and four-wave mixing. Here we briefly describe Rayleigh and Raman scattering; nonlinear methods are beyond the scope of this text.

The Rayleigh and Raman scattering processes are illustrated in Fig. 13.10. Rayleigh scattering involves the excitation of the atom or molecule, in which energy is added but the atom/molecule relaxes back to the original energy state. Raman scattering is similar, except that the atom/molecule relaxes to a state of higher or lower energy. These two

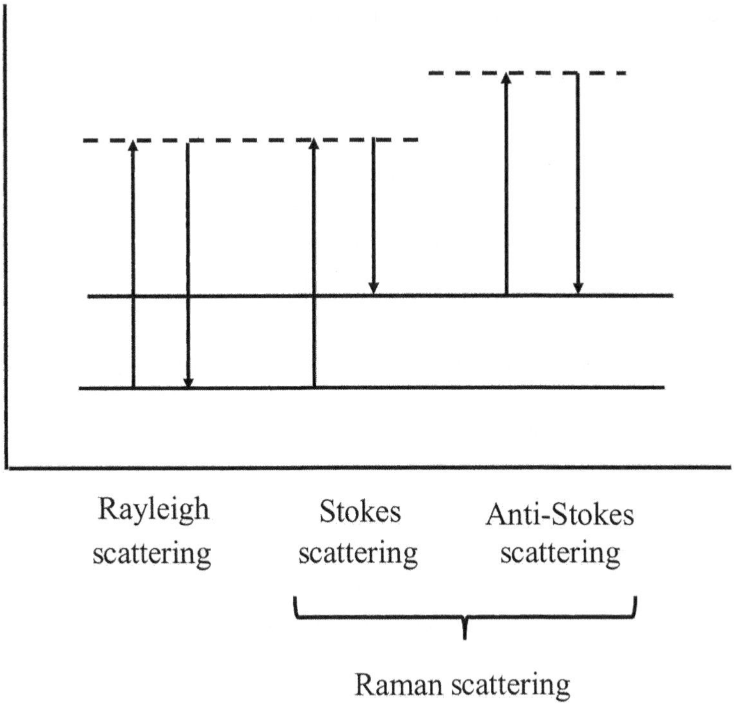

Figure 13.10 Rayleigh and Raman scattering processes.

situations are called, respectively, Stokes or anti-Stokes scattering. The scattered light can result in a change in either rotational or vibrational state. The dashed lines represent "virtual" states in that they don't correspond to a stationary quantum state.

While detailed theory is beyond the scope of this book, one can understand Rayleigh and Raman scattering using classical electromagnetic theory, which states that an electromagnetic field can be generated by oscillating charge. Atoms and molecules, of course, are composed of positive and negative charges. If they are placed in an oscillating electric field, then a second field can be generated. The measure of charge separation induced by an applied field is called the induced dipole moment, and the relation between the dipole moment $\vec{p}$ and the applied field is given by the polarizability α:

$$\vec{p} = \alpha \epsilon_0 \vec{E} \tag{13.63}$$

The polarizability can be expanded about a particular coordinate of motion, for example, a vibrational coordinate Q:

$$\alpha = \alpha_o + \left(\frac{\partial \alpha}{\partial Q} \right)_0 Q \tag{13.64}$$

If Q is oscillating at the vibrational frequency of the molecule ω_v, then

$$Q = Q_0 \cos \omega_v t \tag{13.65}$$

As a result, the induced dipole moment becomes

$$\vec{p} = \alpha_0 \vec{E}_0 \cos \omega_0 t + \epsilon_0 \frac{Q_0 \vec{E}_0}{2} \left(\frac{\partial \alpha}{\partial Q} \right)_0 \left[\cos(\omega_0 - \omega_v)t + \cos(\omega_0 + \omega_v)t \right] \quad (13.66)$$

The induced dipole moment contains the original excitation frequency, plus frequencies at $\omega_0 \pm \omega_v$. Note, however, that a molecule has multiple vibrational levels, each spaced approximately ω_v apart. This leads to a series of Stokes and anti-Stokes frequencies, as illustrated in Fig. 13.10.

13.8 Summary

In this chapter we explored the subject of spectroscopy, which is the use of light interacting with matter to study the state of the matter.

13.8.1 The Absorption and Emission of Radiation

Here we discussed the Einstein view of the absorption and emission of radiation, as illustrated in Fig. 13.11, where the rates of absorption and emission W are a function of the Einstein coefficients B_{12}, B_{21}, and A_{21}. These coefficients are atomic/molecular properties and once measured or obtained from theory, are known. The overall rates depend on other parameters. For example,

$$W_{21} = \int_0^\infty B_{21} E_\nu(\nu) \phi(\nu) d\nu \quad (13.67)$$

where $E_\nu(\nu)$ is the spectral irradiance and $\phi(\nu)$ the normalized spectral line shape function.

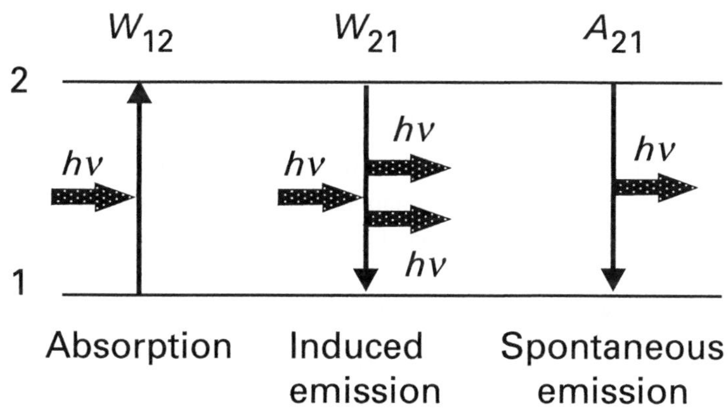

Figure 13.11 The Einstein radiative processes.

13.8.2 Spectral Line Broadening

When the spectral nature of light that has been absorbed or emitted by a particular atom or molecule is examined in detail, it is found that the light is absorbed or emitted over a finite bandwidth, contrary to what would be expected from a simple quantum-mechanical point of view. This effect is called line broadening. Under normal conditions, the important causes of broadening are the Doppler effect and pressure broadening:

$$\phi(\nu) = \left(\frac{mc^2}{2\pi kT\nu_0^2}\right) \exp\left[\frac{-mc^2(\nu - \nu_0)}{2kT\nu_0^2}\right] \tag{13.68}$$

$$\phi(\nu) = \frac{1}{\pi}\frac{\Delta\nu_L/2}{(\nu - \nu_0)^2 + (\Delta\nu_L/2)^2} \tag{13.69}$$

13.8.3 Spectral Transitions

When light is absorbed, the result is a change in the quantum state of the atom or molecule. For atoms, only changes in electronic state are allowed, while for molecules, changes in rotational, vibrational, and electronic states are possible.

13.8.4 Types of Spectroscopies

There are many types of spectroscopies. We discussed three: absorption and emission, laser-induced fluorescence, and Rayleigh and Raman spectroscopy. See Figs. 13.1, 13.9, and 13.10.

13.9 Problems

13.1 An atomic lamp is often used as a calibration source for spectrometers because the spectral emission lines are well defined and spaced such that they are easy to identify. An electric current is used to excite the valence electrons to excited states from which they decay to the ground state emitting photons. For the hydrogen atom, calculate the frequency and wavelength at which the first five lines in increasing wavelength would appear.

13.2 This problem illustrates both the structure of a diatomic molecule and the large disparity between changes in rotational, vibrational, and electronic energy levels. It also illustrates the formation of a band head, the abrupt end of a spectroscopic band.

The following data for I_2 are available from the NIST website:

$X^1\Sigma_g^+$	$T_e = 0\ \text{cm}^{-1}$	$\omega_e = 214.50\ \text{cm}^{-1}$	$\omega_e x_e = 0.614\ \text{cm}^{-1}$
	$B_e = 0.03737\ \text{cm}^{-1}$	$\alpha_e = 0.000113\ \text{cm}^{-1}$	$D_e = 4.2 \times 10^{-9}\ \text{cm}^{-1}$
$B^3\Pi_{0+u}$	$T_e = 15769.01\ \text{cm}^{-1}$	$\omega_e = 125.69\ \text{cm}^{-1}$	$\omega_e x_e = 0.764\ \text{cm}^{-1}$
	$B_e = 0.02903\ \text{cm}^{-1}$	$\alpha_e = 0.000158\ \text{cm}^{-1}$	$D_e = 5.4 \times 10^{-9}\ \text{cm}^{-1}$

(a) Construct an energy-level diagram for the ground electronic state. Consider only the first 10 rotational levels within the first two vibrational levels. Label the ordinate in cm^{-1} and indicate the total degeneracy of each individual rotational energy level on the diagram.

(b) Determine the energies (in cm^{-1}) of the 10 purely rotational transitions for the $v = 0$ energy level within the ground electronic state and plot them on a second energy-level diagram. Draw a vertical line representing $v = 0$, $J = 10 \rightarrow 9$ on the diagram from part (a).

(c) Calculate the energies (in cm^{-1}) for the electronic transitions $P(1)$ through $P(4)$ and $R(0)$ through $R(5)$ and plot them on a third diagram. Make the abscissa the transition energy, and the ordinate the line intensity. Assume the line intensity scales with the Boltzmann ratio for the lower state.

14 Chemical Kinetics

Another area of non-equilibrium behavior is chemical kinetics, important in fields like combustion, chemical vapor deposition, and atmospheric chemistry. Kinetics, or the rate at which chemical reactions occur, can have a large effect on the behavior of systems, determining the degree of pollutant formation in combustion, the composition of solid material formed in deposition processes, the concentration of ozone in the atmosphere, and many other applications. And while it is beyond the scope of this book to explore the subject too deeply, we will briefly treat the physics of reactions and reaction rates for gas-phase reactions. Further resources include Steinfeld and Francisco [56], Marin and Yablonsky [57], and Houston [58].

A chemical reaction is one in which two or more species collide with each other, resulting in a rearrangement of the atoms that make up the original species. An example is the abstraction of a hydrogen atom from methane by collision with another hydrogen atom:

$$CH_4 + H \rightarrow CH_3 + H_2 \tag{14.1}$$

This reaction is called "elementary" because it can actually occur as written. Often, complex systems of reactions are expressed "globally." If you studied combustion reactions in undergraduate thermodynamics, you will recognize the global oxidation reaction equation of methane:

$$CH_4 + 2O_2 \rightarrow CO_2 + 2H_2O \tag{14.2}$$

In fact, the oxidation of methane takes place as a result of a large number of elementary reactions. The global expression is a statement of conservation of atoms as the reaction proceeds from reactants to products. In this case, one carbon in the reactants produces one in the products. Likewise, four atoms each for oxygen and hydrogen.

14.1 Reaction Rate

Consider a general reaction

$$aA + bB \rightarrow cC + dD \tag{14.3}$$

The rate at which this reaction occurs can be written as

$$R = -\frac{1}{a}\frac{d[A]}{dt} = -\frac{1}{b}\frac{d[B]}{dt} = \frac{1}{c}\frac{d[C]}{dt} = \frac{1}{d}\frac{d[D]}{dt} \tag{14.4}$$

The square bracket notation, [A] for example, indicates concentration of the given species.

From our discussions in Chapter 12, we know that the collision frequency between two particles is proportional to the product of their number densities or concentrations. Therefore, for the general reaction above with two reactants,

$$R = k[\mathrm{A}]^a[\mathrm{B}]^b \tag{14.5}$$

where the details of the collisions that determine whether a reaction actually occurs are contained in k, the reaction rate constant. Equation (14.5) is called a rate equation and is a first-order ordinary differential equation. In the fully general case, it can be written as

$$R = k\prod_i c_i^{n_i} \tag{14.6}$$

n_i is called the "order" with respect to the concentration of component c_i. For example, for a three-body reaction such as

$$\mathrm{H} + \mathrm{H} + \mathrm{H_2O} \rightarrow \mathrm{H_2} + \mathrm{H_2O} \tag{14.7}$$

the rate becomes

$$R = k[\mathrm{H}]^2[\mathrm{H_2O}] \tag{14.8}$$

The term "molecularity" refers to the number of reactants involved. This is illustrated by the following reactions:

- Unimolecular: $C_2H_2 \rightarrow CH_2 + C$
- Bimolecular: $O + H_2 \rightarrow OH + H$
- Trimolecular: $CO + O + H_2O \rightarrow CO_2 + H_2O$

Higher-order reactions are not seen in practice.

For simplicity of use, and based on experimental data, complex sets of reactions are often expressed as global reactions. An example is the oxidation of propane, which can be expressed as

$$R = 8.6 \times 10^{11} \exp\left(-\frac{30}{R_u T}\right)[\mathrm{C_3H_8}]^{0.1}[\mathrm{O_2}]^{1.65} \tag{14.9}$$

Note that for a global reaction the order of each reactant can be a non-integer. This would not be the case for an elementary reaction.

14.2 Reaction Rate Constant and the Arrhenius Form

The reaction rate is determined by the rate at which reactants collide times the probability that a reaction will occur. Experiment shows that many rate constants can be described by a function of the form

$$k = AT^n \exp(-E_A/RT) \tag{14.10}$$

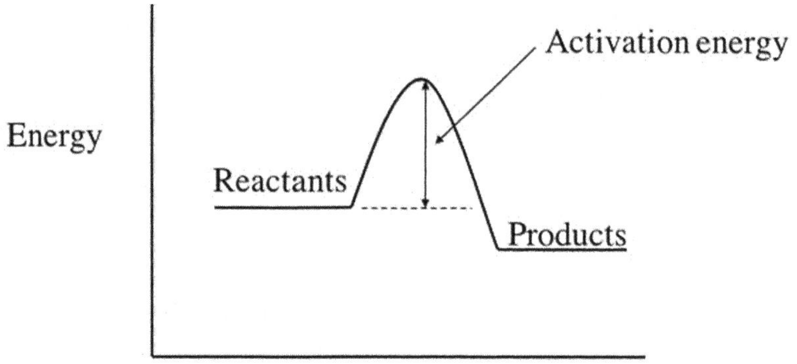

Figure 14.1 Potential energy barrier.

In this form, originally derived by Arrhenius [59], A is called the pre-exponential or frequency factor and E_A is the activation energy. The T^n term accounts for the temperature dependence of the pre-exponential factor, while E_A is approximately the collision energy for a reaction to occur. This is illustrated in Fig. 14.1, a reaction path energy diagram. As we learned earlier, molecules are held together by electrostatic forces. For a reaction to occur, the dissociation energy must be overcome. The plot shows a reaction where the products lie at a lower energy than the reactants. However, for the atomic rearrangement to occur, an energy barrier must be crossed. If there was no barrier, the activation energy would be zero.

Recall from Chapter 12 that the rate of collisions between two species A and B can be written as

$$\frac{\text{\# collisions all A with all B}}{\text{unit time}} = \pi d^2 v_r n_A n_B \tag{14.11}$$

We can incorporate some reaction dynamics into πd^2 if we define it as an *effective* collision cross-section, such that the rate of a given process is

$$\frac{dn_A}{dt} = \int \pi d^2 v_r n_A n_B d\{\text{all } v_r \text{ and orientations}\} \tag{14.12}$$

Factoring a rate coefficient out of this expression, we obtain

$$k = \int \pi d^2 v_r d\{\text{all } v_r \text{ and orientations}\} \tag{14.13}$$

If we only consider spherically symmetric atoms or molecules, then this expression becomes

$$k = \int \pi d^2 v_r dv_r \tag{14.14}$$

Now, suppose we consider the dissociation reaction

$$\mathrm{A_2 + M \rightarrow A + A + M} \tag{14.15}$$

where M in this case represents any collision partner. We would expect the probability of reaction to be related to the collision energy, as shown in Fig. 14.2. The maximum amount of energy available in a given collision is

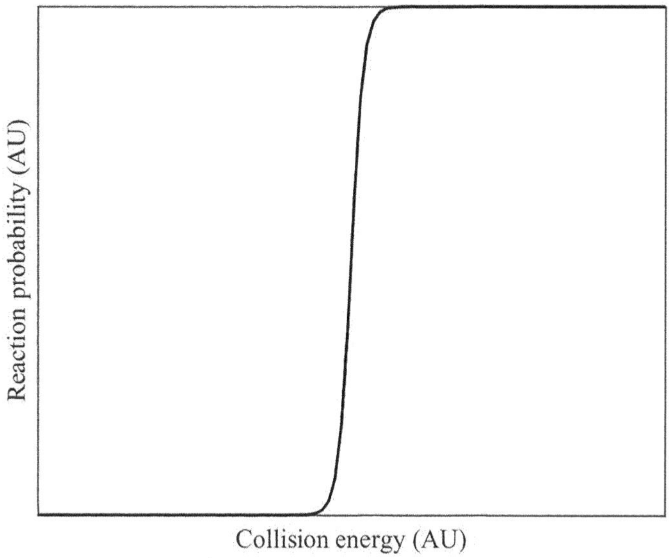

Figure 14.2 Reaction probability as a function of collision energy.

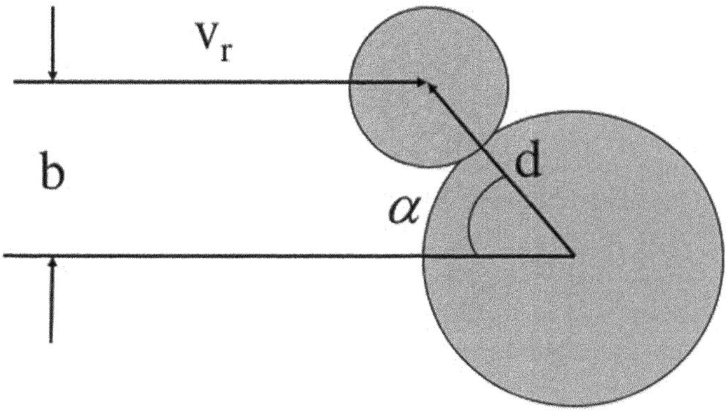

Figure 14.3 Collision geometry.

$$\epsilon_T = \frac{1}{2}\mu v_r^2 \tag{14.16}$$

where $\mu = \frac{m_A m_M}{m_A + m_M}$ is the reduced mass and v_r is the relative collision velocity. However, this amount of energy is only available if the molecules collide along the line of centers. As illustrated in Fig. 14.3, particles can collide with an offset from the line of centers. The offset, b, is called the impact factor.

The relationship between the relative velocity and velocity along the line of centers is then

$$v_{lc} = v_r \cos\alpha = v_r \left[(d^2 - b^2) \right]^{1/2} \tag{14.17}$$

Thus,

$$\epsilon_{lc} = \epsilon_T(1 - b^2/d^2) \tag{14.18}$$

For a reaction to occur, we must have

$$\epsilon_{lc} \geq \epsilon_T^0 \tag{14.19}$$

where ϵ_T^0 is a threshold energy about equal to the dissociation energy. The equation

$$\epsilon_{lc} = \epsilon_T^0 = \epsilon_T(1 - b^2/d^2) \tag{14.20}$$

defines, for a given v_{rel}, the maximum impact parameter at which reaction can occur. Solving for b_{max},

$$b_{max} = d(1 - \epsilon_T^0/\epsilon_T)^{1/2} \tag{14.21}$$

An effective reaction cross-section can be defined as

$$\sigma = \pi d^2(1 - \epsilon_T^0/\epsilon_T) \tag{14.22}$$

One can show (Vincenti and Kruger [2]) that in center of mass coordinates, the number of collisions per unit time and volume, with b between b and $b+db$, and relative velocity between v_r and $v_r + dv_r$, is

$$dZ = n_A n_B \left(\frac{\mu}{kT}\right)^{3/2} \left(\frac{2}{\pi}\right)^{1/2} 2\pi b db v_r^3 e^{-\mu v_r^2/2kT} dv_r \tag{14.23}$$

To obtain the total number of collisions we must integrate over all allowed impact parameters and relative velocities. First, integrating over the impact factor

$$\int_0^{b_{max}} 2\pi b db = \pi b_{max}^2 \tag{14.24}$$

For the next step, it is easier to transform velocity to kinetic energy:

$$\epsilon_T = \frac{1}{2}\mu v_r^2 \tag{14.25}$$

so that

$$d\epsilon_T = \mu v_r dv_r \tag{14.26}$$

Therefore,

$$v_r^3 e^{-\mu v_r^2/2kT} dv_r = \frac{v_r^2}{\mu} e^{-\epsilon_T/kT} d\epsilon_T = \frac{2}{\mu^2} \epsilon_T e^{-\epsilon_T/kT} d\epsilon_T \tag{14.27}$$

The rate constant is then

$$k(T) = \left(\frac{1}{\pi \mu}\right)^{1/2} \left(\frac{2}{kT}\right)^{3/2} \int_{\epsilon_T^0}^{\infty} \pi d^2(1 - \epsilon_T^0/\epsilon_T)\epsilon_T e^{-\epsilon_T/kT} d\epsilon_T \tag{14.28}$$

$$k(T) = \pi d^2 \bar{v}_r e^{-\epsilon_T^0/kT} \tag{14.29}$$

where

$$\bar{v}_r = \left(\frac{8kT}{\pi\mu}\right)^{1/2} \tag{14.30}$$

This is why we see the rate constant expressed in the Arrhenius form as in Eq. (14.10).

14.2.1 Unimolecular Reactions

Unimolecular reactions present a special case. Consider the reaction

$$\mathrm{A} + \mathrm{M} \rightarrow \mathrm{Products} + \mathrm{M} \tag{14.31}$$

where M is a collision partner that retains its identity following the collision. (It may, however, form a chemical complex with A during the collision.) Obviously, this is a bimolecular reaction, in which the collision of A with M provides energy to dissociate A. Because M remains after the collision, it is convenient to write this as

$$\mathrm{A} \rightarrow \mathrm{products} \tag{14.32}$$

One can then write the rate of reaction as

$$\frac{d[\mathrm{A}]}{dt} = -k_{uni}[\mathrm{A}] \tag{14.33}$$

where the influence of M is contained in the unimolecular rate constant, k_{uni}. The overall rate will be affected by the concentration of M, and through the ideal gas law the pressure. As a result,

$$k_{uni} = f(T, p) \tag{14.34}$$

The reaction takes place as a result of the collision forcing the reactant into an excited quantum state, from which it can decompose. Unimolecular reactions were studied by Lindeman and Hinshelwood (see [56]), who developed a simple analysis based on assuming that the reaction took place in two steps. First, the collision forces the reactant into an excited state which is in quasi-equilibrium with the initial state of the reactant:

$$\mathrm{A} + \mathrm{M} \underset{k_{-1}}{\overset{k_1}{\rightleftarrows}} \mathrm{A}^* + \mathrm{M} \tag{14.35}$$

followed by spontaneous dissociation:

$$\mathrm{A}^* \xrightarrow{k_2} \mathrm{products} \tag{14.36}$$

If one assumes that k_2 is much greater than k_1, the consequence is that the concentration of A^* remains very small during the reaction. If that is the case, then to a reasonable approximation we can write

$$\frac{d[\mathrm{A}^*]}{dt} = k_1[\mathrm{A}] - k_2[\mathrm{A}^*] \sim 0 \tag{14.37}$$

This is called the steady-state approximation, and results in the following relation:

$$-\frac{d[\mathrm{A}]}{dt} = k_{uni}[\mathrm{A}] = k_2[\mathrm{A}^*] = \frac{k_1 k_2[\mathrm{A}][\mathrm{M}]}{k_{-1}[\mathrm{M}] + k_2} \tag{14.38}$$

so that

$$k_{uni} = \frac{k_1 k_2[\mathrm{M}]}{k_{-1}[\mathrm{M}] + k_2} \tag{14.39}$$

For [M] $\rightarrow \infty$, the rate becomes

$$k_\infty = \frac{k_1 k_2}{k_{-1}} \tag{14.40}$$

and for [M] $\rightarrow 0$, it becomes

$$k_0 = k_1[\mathrm{M}] \tag{14.41}$$

Thus, the unimolecular reaction rate can be written as

$$k_{uni} = \frac{k_\infty}{1 + \frac{k_\infty}{k_1[\mathrm{M}]}} = k_\infty \left(\frac{P_r}{1 + P_r} \right) \tag{14.42}$$

where $P_r = k_1[\mathrm{M}]/k_\infty$.

It has been shown that this expression does not do an adequate job of representing the pressure dependency. As a result, it is common to include an additional factor F, as

$$k_{uni} = k_\infty \left(\frac{P_r}{1 + P_r} \right) F \tag{14.43}$$

with the factor F commonly given by the Troe (Gilbert and Smith [60]) form:

$$\log F = \left[1 + \left[\frac{\log P_r + c}{n - d(\log P_r + c)} \right]^2 \right]^{-1} \log F_{cent} \tag{14.44}$$

where

$$F_{cent} = 0.138 e^{670/T} \tag{14.45}$$

$$c = -0.4 - 0.67 \log F_{cent} \tag{14.46}$$

$$n = 0.75 - 1.27 \log F_{cent} \tag{14.47}$$

$$d = 0.14 \tag{14.48}$$

14.2.2 Example 14.1

Plot the unimolecular rate constant at 1500 K for acetaldehyde, a reactive intermediate in both combustion and atmospheric chemistry.

For acetaldehyde, rates are available from Sivaramakrishnan, Michael, and Klippenstein [61] and Harding, Georgievskii, and Klippenstein [62]. They fit their theoretical overall unimolecular decomposition rates using the Troe procedure. The theoretical results were

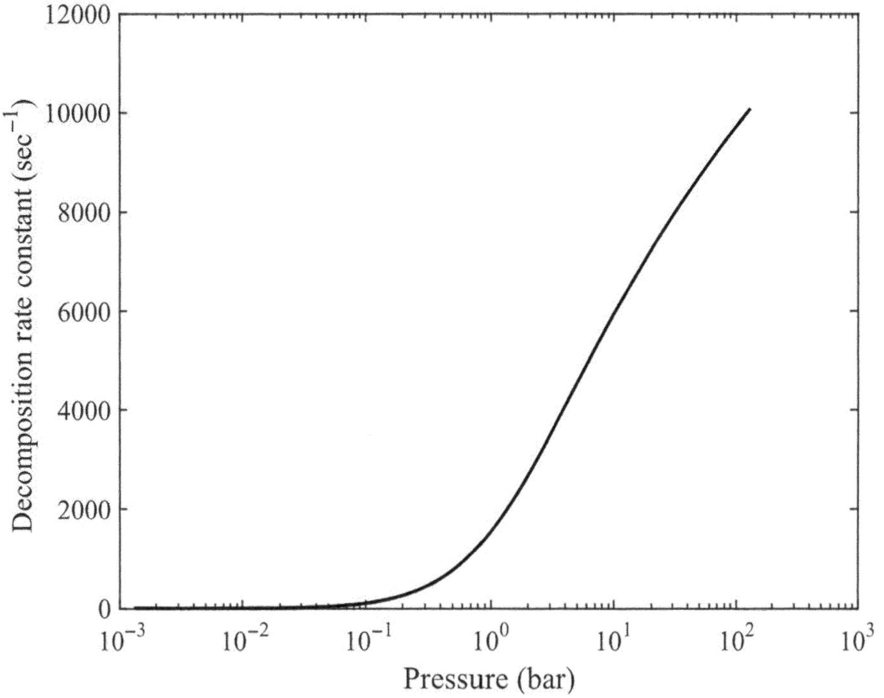

Figure 14.4 Acetaldehyde unimolecular decomposition rate constant k_{uni} at 1500 K.

in fair agreement with the experimental measurements. The fit parameters (valid for 600–2500 K) are given by

$$k_0 = 1.144e59 * (T^{-11.3}) * \exp(-48270/T)\frac{\text{cm}^3}{\text{molec sec}} \tag{14.49}$$

$$k_\infty = 2.72 \times 10^{22} T^{-1.74} e^{-43460/T} \frac{1}{\text{sec}} \tag{14.50}$$

Note that k_0 given in Sivaramakrishnan et al. is in units of cm^3/molecule sec.

Using these rate expressions we can plot k_{uni} as a function of the pressure. Figure 14.4 shows the result for 1500 K.

14.3 More on Reaction Rates

Only if a variety of conditions are met will reaction occur:

- there must be enough energy available
- the molecules must be appropriately oriented
- reaction must be quantum-mechanically allowed
- mass, momentum, and energy must be conserved.

We discussed the first requirement in Section 14.2 for the case of spherically symmetric particles. This led to the Arrhenius form for the rate constant. To take into account molecular structure and reaction processes requires exploring potential energy surfaces (PESs) in detail. If the PES is known, then in principle we can explore all the possibilities by carrying out trajectory calculations and averaging the results over all relative velocities, impact parameters, internal states, and orientations. However, trajectory calculations are very time consuming and large numbers of initial conditions are needed for good statistics. An examination of most PESs will reveal that most trajectories will not result in reaction.

As a result, other methodologies have been pursued, including various versions of transition state theory (TST) and statistical theories.

14.3.1 Transition State Theory

TST assumes the existence of a transition state complex. This is a transitional structure formed during a collision:

$$A + B \rightleftarrows [AB]^* \rightarrow \text{products} \tag{14.51}$$

In TST, the transition state is assumed to be in quasi-equilibrium with the reactants. The reaction rate is the rate at which the energy barrier is crossed.

The rate constant becomes

$$k(T) = \frac{kT}{h} \frac{Q^*}{Q_A Q_B} \exp(-E_0/kT) \tag{14.52}$$

where Q^* is the partition function for the complex. Introducing the standard molar Gibbs free energy

$$\Delta G^{*0} = -RT \ln K_c^* = \Delta H^{\ddagger 0} - T\Delta S^{\ddagger 0} \tag{14.53}$$

the rate constant becomes

$$k(T) = \frac{kT}{h} e^{\Delta S^{\ddagger 0}/R} e^{-\Delta H^{\ddagger 0}/RT} \tag{14.54}$$

If quality quantum-mechanical predictions of the reactants, products, and transition state structures are available, $\Delta S^{\ddagger 0}$ and $H^{\ddagger 0}$ can be calculated. Alternatively, the rate expression can be used to determine the transition state properties by fitting to experimental data.

14.3.2 Statistical Theories: RRKM

As we have learned, at temperatures above absolute zero, molecules assume a distribution of quantum states. As a result, not all reaction collisions start with the same quantum states. Statistical theories account for the quantum state distributions in the reactants and transition states. The theory of Rice, Ramsperger, Kassel, and Marcus (RRKM) is the basis of most modern calculations.

In the context of a unimolecular reaction, to take internal energy into account, k_2 must become energy dependent:

$$\mathrm{A} + \mathrm{M} \overset{dk_1}{\rightarrow} \mathrm{A}^*(E, E + dE) + \mathrm{M} \tag{14.55}$$

$$\mathrm{A}^*(E, E + dE) + \mathrm{M} \overset{k_1}{\rightarrow} \mathrm{A} + \mathrm{M} \tag{14.56}$$

$$\mathrm{A}^*(E, E + dE) \overset{k_2(E)}{\rightarrow} \text{products} \tag{14.57}$$

Applying the steady-state approximation to A* by assuming that A* is in quasi-equilibrium with A, one obtains for the rate of formation of products

$$\frac{d\mathrm{Prod}(E)}{dt} = \frac{k_2(E)dk_1(E)[\mathrm{A}][\mathrm{M}]}{k_{-1}[\mathrm{M}] + k_2(E)} \tag{14.58}$$

Thus,

$$dk_{uni}(E, E + dE) = \frac{k_2(E)dk_1(E)[\mathrm{M}]}{k_{-1}[\mathrm{M}] + k_2(E)} \tag{14.59}$$

and integrating to obtain the total rate constant

$$k_{uni} = \int_{E_0}^{\infty} \frac{k_2(E)dk_1(E)[\mathrm{M}]}{k_{-1}[\mathrm{M}] + k_2(E)} \tag{14.60}$$

where E_0 is the minimum energy for dissociation to occur.

It is assumed that dk_1/k_{-1} represents the equilibrium probability that A* has energy in the range E to $E+dE$. This can be denoted $P(E)dE$. $k_{-1}[\mathrm{M}]$ is the collision frequency ν_c between A* and the molecules M. As a result, the rate expression can be written

$$k_{uni} = \nu_c \int_{E_0}^{\infty} \frac{k_2(E)P(E)dE}{k_2(E) + \nu_c} \tag{14.61}$$

The next step is to determine $k_2(E)$. The RRKM theory predicts how the energy in a molecule is distributed among its possible modes of motion, and what the probability is that a molecule will dissociate because the energy concentrates in one vibrational mode. The assumptions are:

1. All internal states of A* at energy E are accessible and will ultimately lead to decomposition.
2. Vibrational energy redistribution within the energized molecule is much faster than unimolecular reaction.

These assumptions imply that the conversion of A to A* is the rate-limiting step. This turns the problem into one of energy transfer. To solve the problems, a set of rate equations that describe the population of individual energy levels, or a range of energies in the case of large molecules, is solved. These are often called "master equations" and take the form

$$\frac{[\mathrm{A}(E_i)]}{dt} = \sum_{j \neq i} k_{ij}[\mathrm{M}] - \sum_{j \neq i} k_{ij}[\mathrm{A}(E_i)] - \sum_{m} k_2(E_i)[\mathrm{A}(E_i)] \tag{14.62}$$

where $[A(E_i)]$ is the concentration of A with energy E to $E + dE$, k_{ij} is the rate constant for energy transfer from state i to j, and $k_2(E_i)$ is the rate constant for dissociation from state i.

The number of master equations required depends on the number of important quantum states of the reactant and transition species. The distribution of quantum states with energy is called the "density of states." Even for relatively small polyatomic molecules, the density of states for internal rotational and vibrational motion can be quite large, many orders of magnitude. As a result, the equations are solved numerically, and a tradeoff between accuracy and computational expense is usually required. Of course, the energy transfer rate must be known to carry out the calculation. There are numerous models in the literature, most famously energy gap expressions that scale the transfer rate with energy difference.

14.4 Reaction Mechanisms

In practice, most reacting systems undergo a large number of elementary reactions as the reactant progresses toward the final products. An example is the oxidation of methane, the major component of natural gas. A commonly used mechanism is GRI-Mech 3.0 [63], which includes 53 species and over 400 reactions.

A simpler example is hydrogen oxidation, which forms the basis of many larger hydrocarbon mechanisms. The global reaction is

$$H_2 + 1/2O_2 \rightarrow H_2O \tag{14.63}$$

One of the standard hydrogen–oxygen mechanisms [13] is shown in Table 14.1. It is composed of 19 reactions divided into four general groupings. The first group are the H_2/O_2 chain reactions. Chain reactions are ones that either increase or maintain the number of highly reactive species. Species that do not have a full complement of covalent bonds tend to be more energetic and react more readily. These are called radicals. Examples are H and O atoms, and OH, the hydroxyl radical. Next come a set of dissociation/recombination reactions. The dissociation reactions, also called initiation reactions, are required to start the cascade of reactions that break apart the reactants, ultimately resulting in equilibrium products. Recombination reactions are ones that either lead back to reactants or to final equilibrium products, H_2O in this case. The third group of reactions involve the hydroperoxy radical, HO_2. Finally, there is a group of reactions involving hydrogen peroxide, H_2O_2.

All these reactions are required to predict the observed behavior of hydrogen/oxygen combustion. This is illustrated in Fig. 14.5, which shows the explosion behavior of H_2/O_2 mixtures as a function of temperature and pressure.

The set of reactions results in a matching set of rate equations, one for each species involved. In the H_2/O_2 example, there are 11 species. In large hydrocarbon mechanisms there can be several hundred species and thousands of reactions. The task of solving the time-dependent rate equations is complicated by the many orders of magnitude range

Table 14.1 Detailed H_2/O_2 reaction mechanism (Li et al. [13])

	Reaction		A	n	E
	H_2/O_2 chain reactions				
1.	$H + O_2 = O + OH$		3.55×10^{15}	−0.41	16.6
2.	$O + H_2 = H + OH$		5.08×10^{4}	2.67	6.29
3.	$H_2 + OH = H_2O + H$		2.16×10^{8}	1.51	3.43
4.	$O + H_2O = OH + OH$		2.97×10^{6}	2.02	13.4
	H_2/O_2 dissociation/recombination reactions				
5.	$H_2 + M = H + H + M$		4.58×10^{19}	−1.40	104.38
	$H_2 + Ar = H + H + Ar$		5.84×10^{18}	−1.10	104.38
	$H_2 + He = H + H + He$		5.84×10^{18}	−1.10	104.38
6.	$O + O + M = O_2 + M$		6.16×10^{15}	−0.50	0.00
	$O + O + Ar = O_2 + Ar$		1.89×10^{13}	0.0	−1.79
	$O + O + He = O_2 + He$		1.89×10^{13}	0.0	−1.79
7.	$O + H + M = OH + M$		4.71×10^{18}	−1.0	0.00
8.	$H + OH + M = H_2O + M$		3.8×10^{22}	−2.00	0.00
	Formation and consumption of HO_2				
9.	$H + O_2 + M = HO_2 + M$	k_0	6.37×10^{20}	−1.72	0.52
	$H + O_2 + M = HO_2 + M$	k_0	9.04×10^{19}	−1.50	0.49
		k_∞	1.48×10^{12}	0.60	0.00
10.	$HO_2 + H = H_2 + O_2$		1.66×10^{13}	0.00	0.82
11.	$HO_2 + H = OH + OH$		7.08×10^{13}	0.00	0.30
12.	$HO_2 + O = OH + O_2$		3.25×10^{13}	0.00	0.00
13.	$HO_2 + OH = H_2O + O_2$		2.89×10^{13}	0.00	−0.50
	Formation and consumption of H_2O_2				
14.	$HO_2 + HO_2 = H_2O_2 + O_2$		4.20×10^{14}	0.00	11.98
	$HO_2 + HO_2 = H_2O_2 + O_2$		1.30×10^{11}	0.00	−1.63
15.	$H_2O_2 + M = OH + OH + M$	k_0	1.20×10^{17}	0.00	45.5
		k_∞	2.95×10^{14}	0.00	48.4
16.	$H_2O_2 + H = H_2O + OH$		2.41×10^{13}	0.00	3.97
17.	$H_2O_2 + H = H_2 + HO_2$		4.82×10^{13}	0.00	7.95
18.	$H_2O_2 + O = OH + HO_2$		9.55×10^{6}	2.00	3.97
19.	$H_2O_2 + OH = H_2O + HO_2$		1.00×10^{12}	0.00	0.00
	$H_2O_2 + OH = H_2O + HO_2$		5.8×10^{14}	0.00	9.56

of characteristic rates encountered. This requires the use of stiff ordinary differential equation solvers. Most chemical kinetic codes use detailed balance, as discussed in Section 6.2.6, to calculate reverse rates, and this in turn requires thermodynamic properties for each species, which are typically provided as polynomial fitting coefficients, as described in Section 6.2.2.

There are two programs worth mentioning. CHEMKIN is a commercial code, now available from ANSYS. It was initially written by a group at Sandia National Laborato-

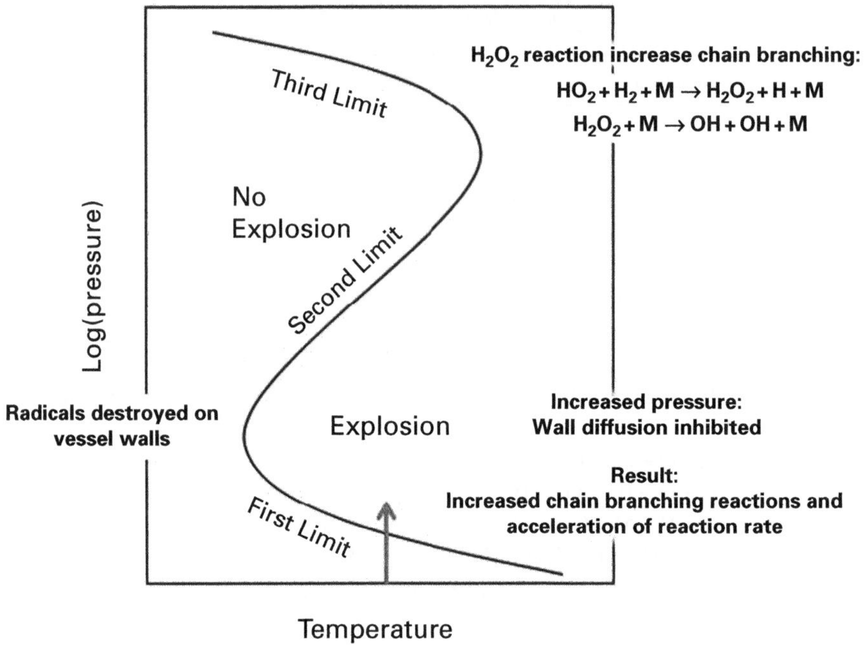

Figure 14.5 H_2/O_2 explosion behavior.

ries and commercialized by a company called Reaction Design. An open source code is Cantera [64], the development of which was initiated by Professor David G. Goodwin at the California Institute of Technology.

14.4.1 Example 14.2

Use either Cantera or CHEMKIN to calculate the time-dependent behavior for a stoichiometric H_2/O_2 mixture at 800 K and 1 atm using the Li et al. [13] mechanism and thermodynamic properties.

The results of the calculation are shown in Figs 14.6 and 14.7. As can be seen, the major species H_2, O_2, and H_2O remain relatively constant, abruptly changing values at about 4.92 sec. This time can be defined as the ignition delay period. However, the radical species H_2O_2, HO_2, H, and O show a much different behavior, rising slowly starting immediately at time zero, then abruptly changing near the ignition delay point. This is because the dissociation reactions for H_2 and O_2 have large activation energies and it requires some time for the radical populations to build to the point where the other reaction kicks in.

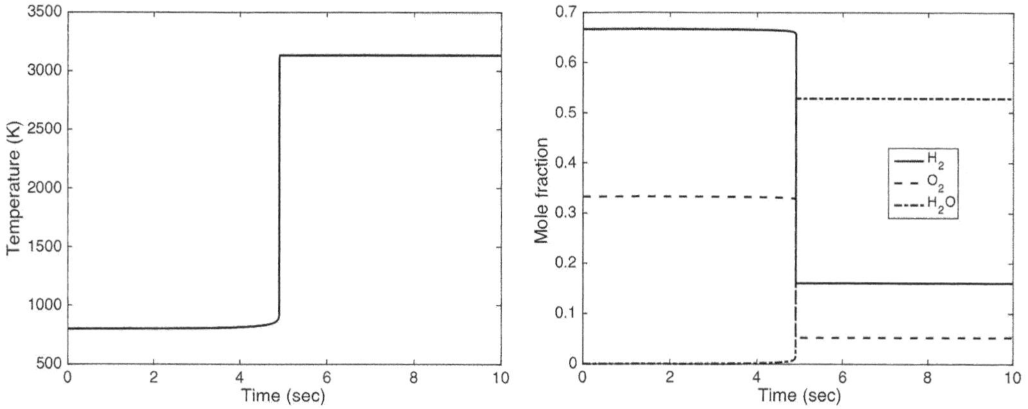

Figure 14.6 Temperature, H_2, O_2, and H_2O vs. time.

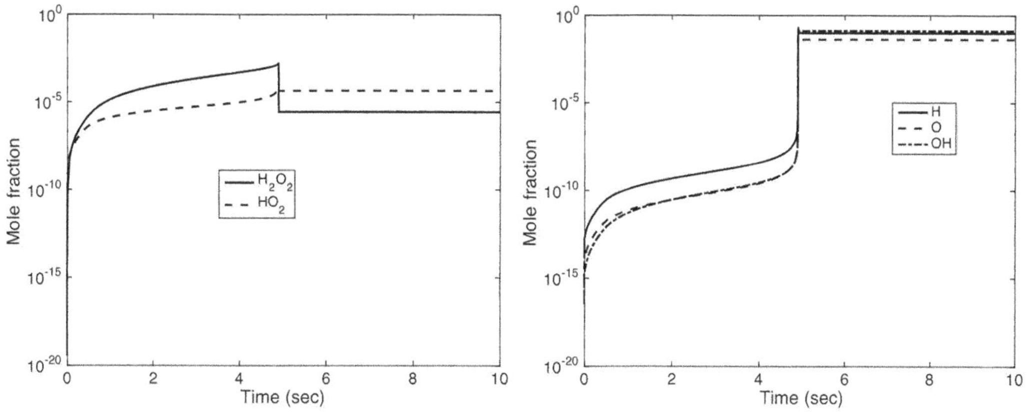

Figure 14.7 H_2O_2, HO_2, H, and O vs. time.

14.5 Summary

In this chapter we explored some aspects of chemical kinetics.

14.5.1 Reaction Rate

Consider a general reaction

$$aA + bB \rightarrow cC + dD \tag{14.64}$$

The rate at which this reaction occurs can be written as

$$R = -\frac{1}{a}\frac{d[A]}{dt} = -\frac{1}{b}\frac{d[B]}{dt} = \frac{1}{c}\frac{d[C]}{dt} = \frac{1}{d}\frac{d[D]}{dt} \tag{14.65}$$

14.5.2 Reaction Rate Constant and the Arrhenius Form

Experiment shows that many rate constants can be described by a function of the form

$$k = AT^n \exp(-E_A/RT) \tag{14.66}$$

14.5.3 Unimolecular Reactions

Unimolecular reactions present a special case. Consider the reaction

$$\mathrm{A{+}M} \rightarrow \text{products} + \mathrm{M} \tag{14.67}$$

This reaction proceeds first by forcing the reactant into an excited state which is in quasi-equilibrium with the initial state of the reactant:

$$\mathrm{A} + \mathrm{M} \underset{k_{-1}}{\overset{k_1}{\rightleftarrows}} \mathrm{A}^* + \mathrm{M} \tag{14.68}$$

followed by spontaneous dissociation

$$\mathrm{A}^* \xrightarrow{k_2} \text{products} \tag{14.69}$$

Applying the steady-state approximation, we showed that

$$k_{uni} = \frac{k_1 k_2[\mathrm{M}]}{k_{-1}[\mathrm{M}] + k_2} \tag{14.70}$$

A more complete expression is

$$k_{uni} = k_\infty \left(\frac{P_r}{1 + P_r}\right) F \tag{14.71}$$

14.5.4 More on Reaction Rates

Here we discussed the requirements for a reaction to occur, and two methods commonly used to estimate reaction rates, transition state theory (TST), and statistical theories, mainly RRKM.

14.5.5 Reaction Mechanisms

In this short section we discussed reaction mechanisms, giving a couple of examples.

14.6 Problems

14.1 Consider the following unimolecular reaction:

$$\mathrm{H_2O_2} + \mathrm{M} \rightarrow \mathrm{OH} + \mathrm{OH} + \mathrm{M}$$

Table 14.2 Rate constant in units of cm^3/molecule sec and temperature in K

Temperature	Rate	Temperature	Rate
300	1.61E-49	1500	7.87E-19
400	6.26E-40	1600	2.37E-18
500	3.56E-34	1700	6.29E-18
600	2.44E-30	1800	1.50E-17
700	1.34E-27	1900	3.25E-17
800	1.52E-25	2000	6.52E-17
900	6.04E-24	2100	1.23E-16
1000	1.15E-22	2200	2.17E-16
1100	1.28E-21	2300	3.67E-16
1200	9.50E-21	2400	5.93E-16
1300	5.20E-20	2500	9.23E-16
1400	2.23E-19		

The high-pressure limit unimolecular rate coefficient for this reaction is

$$k_{uni} = A\exp(-E_a/RT)$$

where $A = 3 \times 10^{14}$ sec^{-1} and $E_a = 305$ kJ/mole.

Calculate the half-life of this reaction at 1 atm and 1000 K.

14.2 Consider the reaction

$$O_2 + OH \rightarrow HO_2 + O$$

Data for the rate coefficient is given in Table 14.2. Fitting the data to the Arrhenius rate form, determine the pre-exponential term in units of cm^3/molecule sec and the activation energy in units of kJ.

14.3 Run a constant-pressure reactor simulation of stoichiometric methane combustion using the GRI-Mech 3.0 [63] mechanism over the temperature range of 1200–2000 K, and plot the NO concentration after 100 ms.

A Physical Constants

Fundamental Physical Constants — Frequently used constants

Quantity	Symbol	Value	Unit	Relative std. uncert. u_r
speed of light in vacuum	c, c_0	299 792 458	m s^{-1}	exact
magnetic constant	μ_0	$4\pi \times 10^{-7}$	N A^{-2}	
		$= 12.566\,370\,614... \times 10^{-7}$	N A^{-2}	exact
electric constant $1/\mu_0 c^2$	ϵ_0	$8.854\,187\,817... \times 10^{-12}$	F m^{-1}	exact
Newtonian constant of gravitation	G	$6.673\,84(80) \times 10^{-11}$	m^3 kg^{-1} s^{-2}	1.2×10^{-4}
Planck constant	h	$6.626\,069\,57(29) \times 10^{-34}$	J s	4.4×10^{-8}
$h/2\pi$	$\hbar$	$1.054\,571\,726(47) \times 10^{-34}$	J s	4.4×10^{-8}
elementary charge	e	$1.602\,176\,565(35) \times 10^{-19}$	C	2.2×10^{-8}
magnetic flux quantum $h/2e$	Φ_0	$2.067\,833\,758(46) \times 10^{-15}$	Wb	2.2×10^{-8}
conductance quantum $2e^2/h$	G_0	$7.748\,091\,7346(25) \times 10^{-5}$	S	3.2×10^{-10}
electron mass	m_e	$9.109\,382\,91(40) \times 10^{-31}$	kg	4.4×10^{-8}
proton mass	m_p	$1.672\,621\,777(74) \times 10^{-27}$	kg	4.4×10^{-8}
proton-electron mass ratio	m_p/m_e	$1836.152\,672\,45(75)$		4.1×10^{-10}
fine-structure constant $e^2/4\pi\epsilon_0\hbar c$	α	$7.297\,352\,5698(24) \times 10^{-3}$		3.2×10^{-10}
inverse fine-structure constant	α^{-1}	$137.035\,999\,074(44)$		3.2×10^{-10}
Rydberg constant $\alpha^2 m_e c/2h$	R_∞	$10\,973\,731.568\,539(55)$	m^{-1}	5.0×10^{-12}
Avogadro constant	N_A, L	$6.022\,141\,29(27) \times 10^{23}$	mol^{-1}	4.4×10^{-8}
Faraday constant $N_A e$	F	$96\,485.3365(21)$	C mol^{-1}	2.2×10^{-8}
molar gas constant	R	$8.314\,4621(75)$	J mol^{-1} K^{-1}	9.1×10^{-7}
Boltzmann constant R/N_A	k	$1.380\,6488(13) \times 10^{-23}$	J K^{-1}	9.1×10^{-7}
Stefan-Boltzmann constant $(\pi^2/60)k^4/\hbar^3 c^2$	σ	$5.670\,373(21) \times 10^{-8}$	W m^{-2} K^{-4}	3.6×10^{-6}
Non-SI units accepted for use with the SI				
electron volt (e/C) J	eV	$1.602\,176\,565(35) \times 10^{-19}$	J	2.2×10^{-8}
(unified) atomic mass unit $\frac{1}{12}m(^{12}\mathrm{C})$	u	$1.660\,538\,921(73) \times 10^{-27}$	kg	4.4×10^{-8}

Source: http://physics.nist.gov/constants.

B Combinatorial Analysis

In carrying out statistical analysis we shall need certain probability formulas. In particular we will need to understand combinations and permutations. Combinations are groups of items without regard to any particular order. For example, "I have a combination of tools in my toolbox." Permutation refers to order, for example, "Wrenches are in the top drawer, screwdrivers in the next drawer, etc." We typically ask questions about combinations and permutations in the following way:

1. How many ways can N distinguishable objects be placed in M different boxes with a limit of one per box? Note that M must be less than or equal to N for there to be only one object per box. The result is

$$\frac{M!}{(M-N)!} \tag{B.1}$$

2. How many ways can N identical, distinguishable objects be placed in M different boxes such that the ith box holds exactly N_i objects?

$$\frac{N!}{\prod_j N_j!} \tag{B.2}$$

3. How many ways can N identical, indistinguishable objects be placed in M different boxes with a limit of one per box?

$$\frac{M!}{N!\,(M-N)!} \tag{B.3}$$

4. How many ways can N identical, indistinguishable objects be placed in M different boxes with no limit on the number per box?

$$\frac{(N+M-1)!}{N!\,(M-1)!} \tag{B.4}$$

C Tables

C.1 Atomic Electron Configurations

Atomic number	Symbol	Configuration	Atomic number	Symbol	Configuration
1	H	$1s^1$	11	Na	$[Ne]3s^1$
2	He	$1s^2$	12	Mg	$[Ne]3s^2$
3	Li	$[He]2s^1$	13	Al	$[Ne]3s^2 3p^1$
4	Be	$[He]2s^2$	14	Si	$[Ne]3s^2 3p^2$
5	B	$[He]2s^2 2p^1$	15	P	$[Ne]3s^2 3p^3$
6	C	$[He]2s^2 2p^2$	16	S	$[Ne]3s^2 3p^4$
7	N	$[He]2s^2 2p^3$	17	Cl	$[Ne]3s^2 3p^5$
8	O	$[He]2s^2 2p^4$	18	Ar	$[Ne]3s^2 3p^6$
9	F	$[He]2s^2 2p^5$	19	K	$[Ar]4s^1$
10	Ne	$[He]2s^2 2p^6$	20	Ca	$[Ar]4s^2$

C.2 Lennard–Jones Parameters

Data are taken from (1) Hippler, Troe, and Wendelken [65] and (2) Mourits and Rummens [66].

Molecule	σ (Å)	ϵ/k (K)	Ref.	Molecule	σ (Å)	ϵ/k (K)	Ref.
He	2.55	10	1	C_2H_6	4.39	234	1
Ne	2.82	32	2	C_3H_8	4.94	275	1
Ar	3.47	114	1	C_4H_{10}	5.40	307	1
Kr	3.66	178	1	C_5H_{12}	5.85	327	1
Xe	4.05	230	1	C_6H_{14}	6.25	343	1
H_2	2.83	60	1	C_7H_{16}	6.65	351	1
D_2	2.73	69	1	C_8H_{18}	7.02	359	1
CO	3.70	105	1	C_9H_{20}	7.34	362	1
N_2	3.74	82	1	$C_{10}H_{22}$	7.72	363	1
NO	3.49	117	1	$C_{11}H_{24}$	8.02	362	1
O_2	3.48	103	1	C_2H_4	4.23	217	1

(Cont)

Molecule	σ (Å)	ϵ/k (K)	Ref.	Molecule	σ (Å)	ϵ/k (K)	Ref.
CO_2	3.94	201	1	C_3H_6	4.78	271	1
N_2O	3.78	249	1	C_2H_2	4.13	224	1
H_2O	2.71	506	1	C_6H_6	5.46	401	1
CH_4	3.79	153	1	C_7H_8	5.92	410	1
C_3H_4	4.667	284.7	2	n-C_8H_{18}	7.024	357.7	2
C_2H_5OH	4.317	450.2	2	CH_3OH	3.657	385.2	2

C.3 Some Properties of Diatomic Molecules

Data are taken from the NIST Chemistry Webbook (webbook.nist.gov), Radzig and Smirov [9], and Ellison [67].

Molecule	ω_e (cm^{-1})	$\omega_e x_e$ (cm^{-1})	B_e (cm^{-1})	α_e (cm^{-1})	T_v (K)	T_r (K)	r_e (Å)	D_e (e.v.)
O_2	1580.2	12.07	1.445	0.0158	2276.6	2.08	1.207	5.11662
N_2	2358.57	14.324	1.99824	0.017318	3395.0	2.88	1.09768	9.76
NO	1904.2	14.075	1.67195	0.0171	2741.0	2.41	1.15077	6.4966
H_2	4401.21	121.33	60.853	3.062	6335.2	87.59	0.74144	4.478
OH	3737.76	84.881	18.910	0.7242	5380.2	27.22	0.96966	5.1014
NH	3282.2	78.3	16.6993	0.6490	4724.5	24.04	1.0362	3.398
CH	2858.5	63.02	14.457	0.534	4114.6	10.81	1.1199	3.468
CO	2169.81358	13.28831	1.93128	0.01750	3123.3	2.78	1.1283	11.9
CN	2068.59	13.087	1.8997	0.0173	2977.6	2.73	1.1718	7.8
C_2	1864.71	13.34	1.8198	0.0176	2684.1	2.76	1.2425	6.2
I_2	214.50	0.614	0.03737	0.000113	308.8	0.0538	2.666	1.542
Cl_2	559.7	2.67	0.2439	0.0014	805.6	0.351	1.987	2.479

D Multicomponent, Reactive Flow Conservation Equations

D.1 The Distribution Function and Flow Properties

We have seen that the thermodynamic properties we observe at the macroscopic level are actually statistical averages over molecular-level behavior. Their values are obtained by integration over the appropriate distribution function. For example, in Chapter 5 we derived the average molecular speed at equilibrium using the Maxwellian speed distribution

$$\bar{c} = \int_0^\infty c f(c) dc = \left(\frac{8kT}{\pi m}\right)^{1/2} \tag{D.1}$$

If c_i is the molecular velocity in the ith direction ($i = x, y$, or z), then the Maxwellian velocity (rather than speed) distribution becomes

$$f(c_1, c_2, c_3) = \left(\frac{m}{2\pi kT}\right)^{3/2} e^{-m(c_1{}^2+c_2{}^2+c_3{}^2)/2kT} \tag{D.2}$$

If we calculate the average speed we would obtain

$$\bar{c_i} = \int_{-\infty}^{\infty} c_i f dc_1 dc_2 dc_3 = 0 \tag{D.3}$$

as expected.

It is usual to define a "thermal" velocity by subtracting the mean velocity from the molecular velocity:

$$C_i \equiv c_i - \bar{c_i} \tag{D.4}$$

A common assumption in fluid mechanics is "local thermodynamic equilibrium" (LTE). Essentially, one assumes that even if there is a mean velocity, the thermal velocity is distributed as though it was in equilibrium. Then the distribution function can be described as a "shifted" Maxwellian:

$$f(c_1, c_2, c_3) = \left(\frac{m}{2\pi kT}\right)^{3/2} e^{-m(\bar{c_1}+C_1)^2+(\bar{c_2}+C_2)^2+(\bar{c_3}+C_3)^2)/2kT} \tag{D.5}$$

In this case,

$$\bar{c_i} = \int_{-\infty}^{\infty} c_i f dc_1 dc_2 dc_3 \neq 0 \tag{D.6}$$

but

$$\bar{C_i} = \int_{-\infty}^{\infty} C_i f dc_1 dc_2 dc_3 = 0 \tag{D.7}$$

Of course, LTE doesn't always apply and in general we must know the form of the distribution function.

Other important properties can also be calculated. These include the fluxes of momentum, energy, and mass.

It turns out that the multicomponent, reactive flow conservation equations can easily be derived based on the velocity distribution function f_k, where the subscript k refers to a particular chemical species. For the given species we assume that

$$f_k(\vec{x}, \vec{c}, t) d\vec{x} d\vec{c} \tag{D.8}$$

gives the probable number of molecules of type k with position between $\vec{x}$ and $\vec{x} + d\vec{x}$ and velocity between $\vec{c}$ and $\vec{c} + d\vec{c}$. Here, $\vec{c}$ is the molecular velocity vector.

The six-dimensional space $(\vec{x}, \vec{c})$ is called *phase space*. Phase space contains both physical and velocity space. The distribution function is thus normalized as the number density, that is

$$\int_{-\infty}^{+\infty} f_k d\vec{c} = n_k \tag{D.9}$$

or

$$\int_{-\infty}^{+\infty} \frac{f_k}{n_k} d\vec{c} = 1 \tag{D.10}$$

Given the distribution function we can calculate the average value of any molecular quantity, Q_k:

$$\bar{Q}_k = \int_{-\infty}^{+\infty} Q_k(\vec{c}) \frac{f_k}{n_k} d\vec{c} \tag{D.11}$$

D.1.1 Conservation Equation Form

The conservation equations can be written in the form of balances. To derive the differential equation forms we carry out a property balance on the differential spatial elements, as illustrated in Fig. D.1.

It is common to collect the density and flux terms on the left side. The joint term is called the substantive or total derivative. Thus,

$$\frac{D}{Dt}(\text{quantity}) = \text{source} \tag{D.12}$$

We first derive forms for the densities and fluxes. We will derive the source terms as we consider each conservation equation.

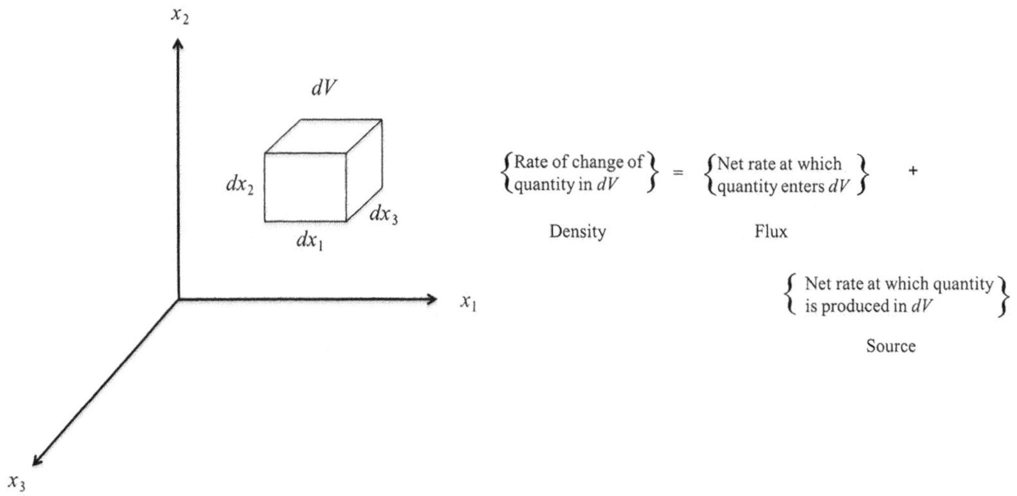

Figure D.1 Conservation balance.

D.1.2 Densities

The velocity distribution function can be used to calculate the appropriate densities (Table D.1). It is conventional to define

$$v_i \equiv \sum_{k=1} N Y_k \bar{c}_{k_i} \tag{D.13}$$

as the mass average velocity, where

$$Y_k \equiv \frac{\rho_k}{\rho} \tag{D.14}$$

is the mass fraction

$$\rho \equiv \sum_{k=1}^{N} \rho_k \tag{D.15}$$

is the total density, and

$$\bar{c}_{k_i} = \iiint_{-\infty}^{+\infty} \frac{c_i}{n_k} f_k dc_1 dc_2 dc_3 \tag{D.16}$$

It is also conventional to define the diffusion velocity as

$$v_{k_i} \equiv \bar{c}_{k_i} - v_i \tag{D.17}$$

and the thermal velocity as

$$C_{k_i} \equiv \bar{c}_{k_i} - v_i \tag{D.18}$$

Note that by the definition of mass average velocity

$$\sum_{k=1}^{N} \rho_k v_{k_i} = 0 \tag{D.19}$$

Table D.1 Forms of the densities

Number density:	$n_k = \int_{-\infty}^{+\infty} f_k d\vec{c}$	
Molar concentration:	$C_k = \frac{n_k}{N_A}$	N_A = Avogadro's number
Mass density:	$\rho_k = m_k n_k$	m_k = molecular mass
Momentum density:	$\int_{-\infty}^{+\infty} \frac{m_k c_i}{n_k} f_k d\vec{c}$	$m_k c_i$ = momentum of a single
	$= m_k \bar{c}_{k_i}$	molecule
Translational energy density:	$\int_{-\infty}^{+\infty} \frac{1}{2n_k} m_k (c_1^2 + c_2^2 + c_3^2) f_k d\vec{c}$	
	$= \frac{1}{2} m_k (\overline{c_1^2} + \overline{c_2^2} + \overline{c_3^2})$	

Also, the total translational energy density becomes

$$\frac{\text{Total K.E.}}{\text{Volume}} = \frac{1}{2}\rho v^2 + \sum_{k=1}^{N} \frac{1}{2}\rho_k \overline{C_k^2} \tag{D.20}$$

where

$$\overline{C_k^2} = \iiint_{-\infty}^{+\infty} (\overline{C_1^2} + \overline{C_2^2} + \overline{C_3^2}) \frac{f_k}{n_k} dc_1 dc_2 dc_3 \tag{D.21}$$

D.1.3 Fluxes

To determine the average fluxes of the various properties, we need to know the flux of molecules across an arbitrary plane. Consider the flux of molecules that strike a surface perpendicular to the x_3 direction. At first consider only those molecules with a given velocity c_i such that the components are in the range c_1 to $c_1 + dc_1$, and so on. The molecules of "class" c_i that will strike an area element ds of the surface in time interval dt are those that, at the beginning of the time interval, are in the slant volume with base area ds, surface parallel to c_1, and of length cdt. As illustrated in Fig. D.2, the volume is $c_3 dtds$ since the height is c_3. The number of molecules of class c_i striking ds in the time interval dt is thus

$$c_3 f_k dc_1 dc_2 dc_3 ds dt \tag{D.22}$$

Thus, the flux of molecules in the class c_i striking the plane is

$$c_3 f_k dc_1 dc_2 dc_3 = c_3 f_k dV_c \tag{D.23}$$

where $dV_c = dc_1 dc_2 dc_3$ is a volume element in velocity space. Therefore, for an arbitrary surface, the flux is

$$J = c_j f_k dV_c \tag{D.24}$$

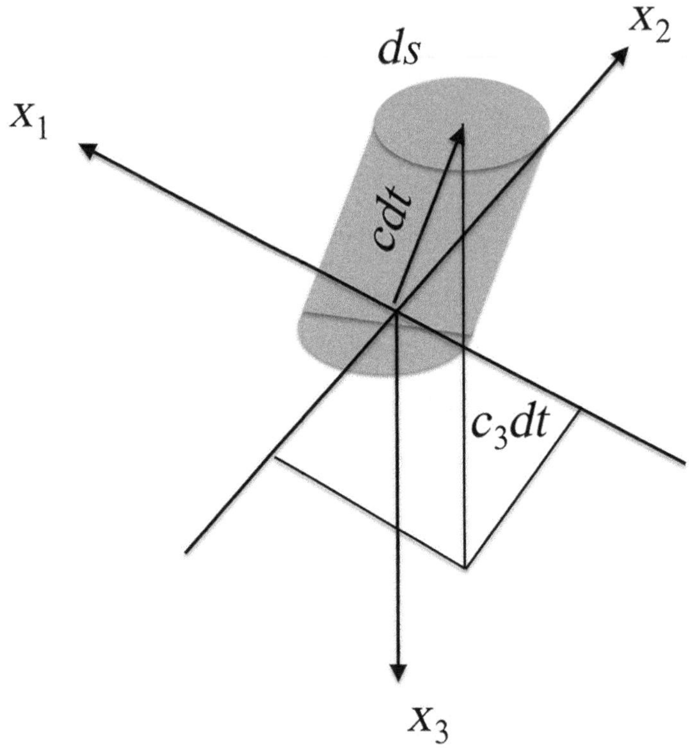

Figure D.2 Geometry for calculating number flux.

where c_j is the velocity normal to the surface. It turns out that when the distribution function for thermal velocity is equal to the Maxwellian,

$$J = \frac{n\bar{c}}{4} \tag{D.25}$$

and this is often a good approximation if the departure from local thermodynamic equilibrium is small.

The mass fluxes then become, for each species,

$$\int_{-\infty}^{+\infty} m_k c_j f_k dV_c = \rho_k \overline{c_{kj}} = \rho_k (v_j + v_{k_j}) \tag{D.26}$$

The total mass flux is

$$\sum_{k=1}^{N} \rho_k \overline{c_{kj}} = \rho v_j \tag{D.27}$$

The momentum flux for species j becomes

$$\int_{-\infty}^{+\infty} m_k c_i c_j f_k dV_c = \rho_k \overline{c_{k_i} c_{k_j}} = \rho_k v_i v_j + \rho_k \overline{C_{k_i} C_{k_j}} \tag{D.28}$$

The total momentum flux is then

$$\rho_k v_i v_j + \sum_{k=1}^{N} \rho_k \overline{C_{k_i} C_{k_j}} \tag{D.29}$$

Finally, the energy flux for each species is

$$\int_{-\infty}^{+\infty} \frac{1}{2} m_k c_k^2 c_{k_j} f_k dV_c = \frac{1}{2} \rho_k \overline{c_k^2 c_{k_j}} \tag{D.30}$$

Now

$$c_{k_i} = C_{k_i} + v_i \tag{D.31}$$

so that

$$c_{k_i}^2 = C_{k_i}^2 + 2C_{k_i} v_i + v_i^2 \tag{D.32}$$

and

$$c_k^2 c_{k_j} = \sum_{k=1}^{N} (C_{k_i}^2 + 2C_{k_i} v_i + v_i^2)(C_{k_j} + v_j) \tag{D.33}$$

Therefore, the flux is

$$\frac{1}{2} \rho_k \left[\overline{c_k^2 c_{k_j}} + 2\overline{C_{k_i} C_{k_j}} v_i + v^2 \overline{C_{k_j}} + v_j \overline{C_k^2} + 2v_j \overline{C_{k_i}} v_i + v_j v^2 \right] \tag{D.34}$$

Now $\overline{C_{k_i}} = v_{k_i}$, so that Eq. (D.34) can be written as

$$\frac{1}{2} \rho_k v_j \left[v^2 + \overline{C_k^2} + 2v_{k_i} v_i \right] + \frac{1}{2} \rho_k \left[\overline{C_k^2 C_{k_j}} + 2v_i \overline{C_{k_i} C_{k_j}} + v^2 v_{k_j} \right] \tag{D.35}$$

Summing over all species, the total energy flux becomes

$$\frac{1}{2} \rho v^2 v_j + \sum_{k=1}^{N} \frac{1}{2} \rho_k \overline{C_k^2} v_j + \sum_{k=1}^{N} \frac{1}{2} \rho_k \left[\overline{C_k^2 C_{k_j}} + 2v_i \overline{C_{k_i} C_{k_j}} \right] \tag{D.36}$$

D.1.4 Conservation Equations

Consider now the differential volume illustrated in Fig. D.3. The mass flow rate for species k entering the volume is equal to the mass flux given by Eq. (D.26) times the surface area of the volume, which in this case is $dx_2 dx_3$ since the flow is in the x_1 direction.

If the volume is small enough, then we would expect that the flow leaving the volume in the x_1 direction would not be greatly different from the inflow. Therefore, expanding the inlet flow about x_1 in a Taylor series and retaining the first-order term should be sufficient. This is illustrated in the figure. Carrying out the same operation in the other two coordinate directions and designating w_k as the source of species k per unit volume and time, a mass balance for the differential volume gives

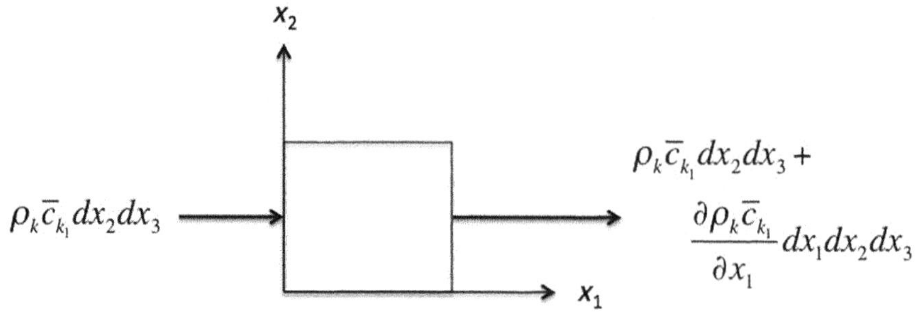

Figure D.3 Mass conservation balance.

$$\begin{aligned}\frac{\partial}{\partial t}(\rho_k dx_1 dx_2 dx_3) = &\left[\rho_k \bar{c}_{k_1} dx_2 dx_3 - \left(\rho_k \bar{c}_{k_1} + \frac{\partial \rho_k \bar{c}_{k_1}}{\partial x_1} dx_1\right) dx_2 dx_3\right] \\ &+ \left[\rho_k \bar{c}_{k_2} dx_1 dx_3 - \left(\rho_k \bar{c}_{k_2} + \frac{\partial \rho_k \bar{c}_{k_2}}{\partial x_1} dx_2\right) dx_1 dx_3\right] \\ &+ \left[\rho_k \bar{c}_{k_3} dx_1 dx_2 - \left(\rho_k \bar{c}_{k_3} + \frac{\partial \rho_k \bar{c}_{k_3}}{\partial x_1} dx_3\right) dx_1 dx_2\right] + w_k dx_1 dx_2 dx_3\end{aligned} \tag{D.37}$$

Gathering and canceling terms, we obtain the species mass conservation equation

$$\frac{\partial \rho_k}{\partial t} = -\frac{\partial \rho_k \bar{c}_{k_1}}{\partial x_1} - \frac{\partial \rho_k \bar{c}_{k_2}}{\partial x_2} - \frac{\partial \rho_k \bar{c}_{k_3}}{\partial x_3} + w_k \tag{D.38}$$

or

$$\frac{\partial \rho_k}{\partial t} + \frac{\partial \rho_k \bar{c}_{k_i}}{\partial x_i} = w_k \tag{D.39}$$

Substituting in the diffusion velocity defined in Eq. (D.17),

$$\frac{\partial \rho_k}{\partial t} + \frac{\partial \rho_k (v_i + v_{k_i})}{\partial x_i} = w_k \tag{D.40}$$

To obtain the total continuity equation, merely sum over all species k. Then

$$\frac{\partial \rho}{\partial t} + \frac{\partial \rho v_i}{\partial x_i} = 0 \tag{D.41}$$

The momentum equation can be derived similarly. Doing so, we obtain

$$\frac{\partial \rho v_i}{\partial t} + \frac{\partial \left[\rho v_i v_j + \sum_{k=1}^{N} \rho_k \overline{C_{k_i} C_{k_j}}\right]}{\partial x_i} = \text{source term} \tag{D.42}$$

A source term for momentum is a body force which acts on individual particles. If $\overline{F}_{k_i}$ is the body force per unit mass of species k acting in the i direction, then the source per unit volume is

$$\sum_{k=1}^{N} \rho_k \overline{F}_{k_i} \tag{D.43}$$

We now explore the meaning of the term

$$\sum_{k=1}^{N} \rho_k \overline{C_{k_i} C_{k_j}} \tag{D.44}$$

Recall that using very simple ideas, and assuming equilibrium, we were able to relate the pressure to the average molecular speed:

$$p = \frac{1}{3} \rho \overline{c^2} \tag{D.45}$$

In the non-equilibrium case we define p as the average of the contributions from the three components of thermal velocity. The partial pressure of species k is then

$$p_k \equiv \frac{1}{3} \rho_k \overline{C_{k_i} C_{k_i}} = \frac{1}{3} \rho_k \overline{C_k^2} \tag{D.46}$$

and the total pressure is

$$p = \sum_{k=1}^{N} p_k \tag{D.47}$$

It is usual to define the viscous stress tensor as the difference between the pressure and the term in Eq. (D.44):

$$\tau_{ij} \equiv -\sum_{k=1}^{N} \left[\rho_k \overline{C_{k_i} C_{k_j}} - p_k \delta_{ij} \right] \tag{D.48}$$

Thus, the i-momentum equation takes on this familiar form:

$$\frac{\partial \rho v_i}{\partial t} + \frac{\partial \rho v_i v_j}{\partial x_i} = -\frac{\partial p}{\partial x_i} + \frac{\partial \tau_{ij}}{\partial x_j} + \sum_{k=1}^{N} \rho_k \overline{F}_{k_i} \tag{D.49}$$

Finally, consider conservation of energy. Performing a similar balance:

$$\frac{\partial}{\partial t} \left\{ \frac{1}{2} \rho v^2 + \sum_{k=1}^{N} \frac{1}{2} \rho_k \overline{c_k^2} \right\} + \frac{\partial}{\partial x_j} \left\{ v_j \left(\frac{1}{2} \rho v^2 + \sum_{k=1}^{N} \frac{1}{2} \rho_k \overline{C_k^2} \right) \right\}$$
$$+ \sum_{k=1}^{N} \frac{1}{2} \rho_k \left\{ \overline{C_k^2 C_{k_j}} + 2 v_i \overline{C_{k_i} C_{k_j}} \right\} = \text{source term} \tag{D.50}$$

The source term for energy arises out of the effect of body forces on individual particles. If F_{k_i} is the body force per unit mass on particles of type k in the i-direction, then

$$\sum_{k=1}^{N} \rho_k \overline{F_{k_j} c_{k_j}} \tag{D.51}$$

is the rate of work per unit volume.

The internal energy due to translation is

$$u_{tr} = \frac{1}{\rho} \sum_{k=1}^{N} \frac{1}{2} \rho_k \overline{C_k^2} \tag{D.52}$$

Thus, the first term in Eq. (D.50) becomes

$$\frac{\partial}{\partial t} \left\{ \rho u_{tr} + \frac{1}{2} \rho v^2 \right\} \tag{D.53}$$

Note also that

$$\sum_{k=1}^{N} \rho_k \overline{C_{k_i} C_{k_j}} = p\delta_{ij} - \tau_{ij} \tag{D.54}$$

Then the second term in Eq. (D.50) becomes

$$\frac{\partial}{\partial x_j} \left\{ \rho v_j \left(u_{tr} + \frac{v^2}{2} \right) + \sum_{k=1}^{N} \frac{1}{2} \rho_k \overline{C_k^2 C_j} + \left(p\delta_{ij} - \tau_{ij} \right) v_i \right\} \tag{D.55}$$

Now, $\delta_{ij} = 0$ if $i \neq j$, so $p\delta_{ij} v_i = p v_i$. Thus, combining the pressure term with the u_{tr} term, this becomes

$$\frac{\partial}{\partial x_j} \left\{ \rho v_j \left(u_{tr} + \frac{p}{\rho} + \frac{v^2}{2} \right) + \sum_{k=1}^{N} \frac{1}{2} \rho_k \overline{C_k^2 C_j} - \tau_{ij} v_i \right\} \tag{D.56}$$

The only term we have not identified is $\sum_{k=1}^{N} \frac{1}{2} \rho_k \overline{C_k^2 C_j}$. This is the only term that is nonzero when $v_j \to 0$, and it carries thermal energy in the j direction. Thus, we define

$$q_j \equiv \sum_{k=1}^{N} \frac{1}{2} \rho_k \overline{C_k^2 C_j} \tag{D.57}$$

as the heat flux vector.

If other forms of internal energy are allowed, then

$$u = u_{tr} + u_{int} \tag{D.58}$$

Defining the enthalpy as

$$h \equiv u + \frac{p}{\rho} \tag{D.59}$$

and

$$q_j \equiv \sum_{k=1}^{N} \frac{1}{2} \rho_k \overline{C_k^2 C_j} + \sum_{k=1}^{N} \overline{\epsilon_{int} C_j} \tag{D.60}$$

we finally obtain

$$\frac{\partial}{\partial t} \left\{ \rho u_{tr} + \frac{1}{2} \rho v^2 \right\} \frac{\partial}{\partial x_j} \left\{ \rho v_j \left(h + \frac{v^2}{2} \right) \right\} = \frac{\partial}{\partial x_j} \left(\tau_{ij} v_i - q_j \right) + \sum_{k=1}^{N} \rho_k \overline{F_{k_j} c_{k_j}} \tag{D.61}$$

It is common to generate a separate equation for the bulk kinetic energy by multiplying the momentum equation by v_i:

$$v_i \left\{ \frac{\partial \rho v_i}{\partial t} + \frac{\partial \rho v_i v_j}{\partial x_i} = -\frac{\partial p}{\partial x_i} + \frac{\partial \tau_{ij}}{\partial x_j} + \sum_{k=1}^{N} \rho_k \overline{F}_{k_i} \right\} \tag{D.62}$$

Now define the total stress tensor as

$$\sigma_{ij} \equiv -p\delta_{ij} - \tau_{ij} \tag{D.63}$$

Substituting this definition and then subtracting from the full energy equation, the thermal energy equation becomes

$$\frac{\partial u}{\partial t} + \rho v_j \frac{\partial u}{\partial x_j} = -\frac{\partial q_i}{\partial x_i} + \sigma_{ij} \frac{\partial v_i}{\partial x_j} + v_i \sum_{k=1}^{N} \rho_k \overline{F_{k_i}} \tag{D.64}$$

Finally, if there is internal heat generation due to radiation, for example, then a term $\dot{Q}$ must be added to the energy equation

$$\frac{\partial u}{\partial t} + \rho v_j \frac{\partial u}{\partial x_j} = -\frac{\partial q_i}{\partial x_i} + \sigma_{ij} \frac{\partial v_i}{\partial x_j} + \dot{Q} + v_i \sum_{k=1}^{N} \rho_k \overline{F_{k_i}} \tag{D.65}$$

D.1.5 Summary

Continuity:

$$\frac{\partial \rho}{\partial t} + \frac{\partial \rho v_i}{\partial x_i} = 0 \tag{D.66}$$

Species:

$$\frac{\partial \rho_k}{\partial t} + \frac{\partial \rho_k (v_i + v_{k_i})}{\partial x_i} = w_k \tag{D.67}$$

Momentum:

$$\frac{\partial \rho v_i}{\partial t} + \frac{\partial \rho v_i v_j}{\partial x_i} = -\frac{\partial p}{\partial x_i} + \frac{\partial \tau_{ij}}{\partial x_j} + \sum_{k=1}^{N} \rho_k \overline{F}_{k_i} \tag{D.68}$$

Thermal energy:

$$\frac{\partial u}{\partial t} + \rho v_j \frac{\partial u}{\partial x_j} = -\frac{\partial q_i}{\partial x_i} + \sigma_{ij} \frac{\partial v_i}{\partial x_j} + \dot{Q} + v_i \sum_{k=1}^{N} \rho_k \overline{F_{k_i}} \tag{D.69}$$

The multicomponent Navier–Stokes equations involve $4N + 19$ unknowns:

Variable	Number of components
ρ_k	N
v_i	3
v_{k_i}	$3N$
ρ	1
τ_{ij}	9
u	1
q_i	3
h	1
p	1
Total	$4N+19$

Meanwhile, the number of equations available are:

Equation	Number of equations	Unknown
Species	N	ρ_k
Continuity	1	ρ
Momentum	3	v_i
Energy	1	u
$h = u + p/\rho$	1	h
$p = \sum_{k=1}^{N} \rho_k RT$	1	p
	$N+7$	

Clearly we have a shortage of equations and must express v_{k_i}, τ_{ij}, and q_i in terms of the other variables. These relationships are called the constitutive or transport relations. Expressions are given in Chapter 12.

E Boltzmann's Equation

To determine the v_{k_i}, τ_{ij}, and q_i requires that we know the distribution function. The distribution function is described by Boltzmann's equation, which we can derive in the same way we derived the conservation equations. Botzmann's equation describes conservation of velocity, that is, it is a continuity equation for velocity class, $c_i \rightarrow c_i + dc_i$.

The density term is merely

$$f_k(\vec{x}, \vec{c}, t) \tag{E.1}$$

However, we must consider this to be density in phase space (Fig. E.1), not just physical space. Thus, we seek the rate of change of the number of molecules in $d\vec{x}$ and $d\vec{c}$, or

$$\frac{\partial}{\partial t}\left[f_k d\vec{x}d\vec{c}\right] = \frac{\partial f_k}{\partial t} d\vec{x}d\vec{c} \tag{E.2}$$

The net flux into $d\vec{x}$ in the 1 direction will be (Fig. E.2)

$$\text{Net}_{\text{in}_1} = -c_1 \frac{\partial f_k}{\partial x_1} d\vec{x}d\vec{c} \tag{E.3}$$

which can be generalized to all three directions as

$$\text{Net}_{\text{in}_j} = -c_j \frac{\partial f_k}{\partial x_j} d\vec{x}d\vec{c} \tag{E.4}$$

Determining the net flux into $d\vec{c}$ is a bit more complex. Molecular velocities can be changed by more than one mechanism. First consider the effect of body forces. If F_{k_i} is the external body force per unit mass, then F_{k_i} equals the acceleration of the molecule. Thus, the net flux into $d\vec{c}$ (Fig. E.3) in the j direction becomes

$$\text{Net}_{\text{in}_j} = \frac{\partial}{\partial c_j}\left[F_{k_j} f_k\right] d\vec{x}d\vec{c} \tag{E.5}$$

Collisions will also change the molecular velocity. We can write the collisional term symbolically as

$$\frac{\partial}{\partial t}\left[f_k d\vec{x}d\vec{c}\right]_{Coll} = \left.\frac{\partial f_k}{\partial t}\right)_{Coll} d\vec{x}d\vec{c} \tag{E.6}$$

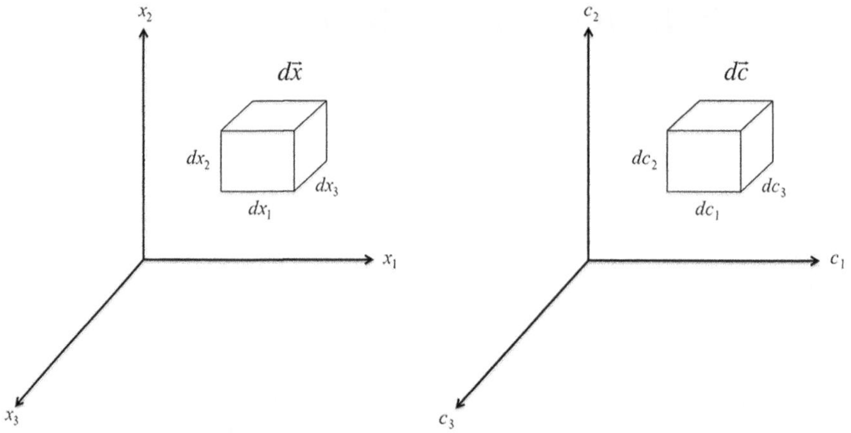

(a) Physical space (b) Velocity space

Figure E.1 Phase space.

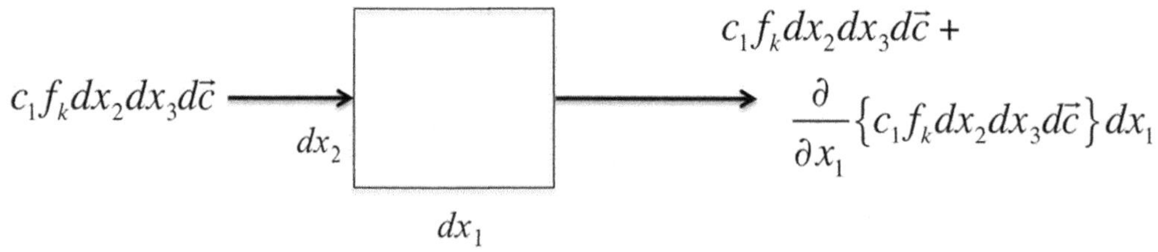

Figure E.2 Flux in physical space.

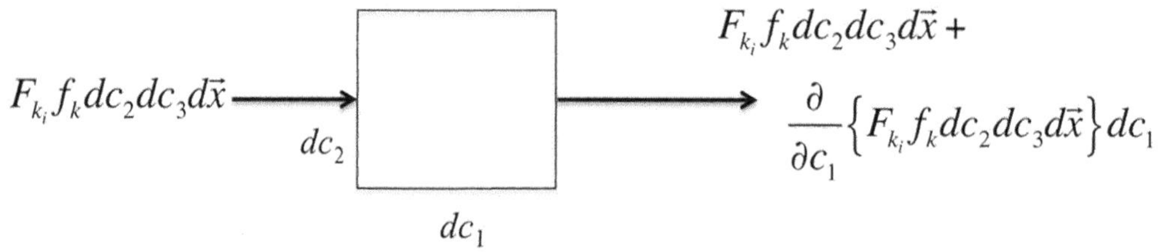

Figure E.3 Flux in velocity space.

One may show that for binary collisions (Hirschfelder, Curtiss, and Bird [14]), this term may be written as

$$\left.\frac{\partial f_k}{\partial t}\right)_{coll} = \sum_{j=1}^{N} \int \int \int (f_k' f_j' - f_k f_j) g_{jk} b\, db\, d\epsilon\, d\vec{c} \tag{E.7}$$

where the integral is over

g_{kj} = reactive collision velocity
b = impact parameter
ϵ = angle of approach out of plane
$\vec{c}$ = velocity of molecule k

and the prime indicates f after the collision.

Putting this all together, we obtain the Boltzmann equation

$$\frac{\partial f_k}{\partial t} + c_j \frac{\partial f_k}{\partial x_j} + \frac{\partial}{\partial x_j}\left[F_{k_j} f_k\right] = \left.\frac{\partial f_k}{\partial t}\right)_{coll} \tag{E.8}$$

Note that the Boltzmann equation is an integro-differential equation and finding solutions is difficult. In particular, the collision integral is very complex and may even involve quantum effects. As a result, many workers have used simplified forms for the collision integral. One of the most common expressions was developed by Bhatngar, Gross, and Krook and is called the BGK model. It is of the form

$$\left.\frac{\partial f_k}{\partial t}\right)_{coll} \simeq \nu_k (f_{k_0} - f_k) \tag{E.9}$$

where ν_k is a collision frequency and f_{k_0} is the equilibrium, or Maxwellian, velocity distribution.

We can estimate the collision frequency as in Chapter 12 by

$$\nu_k = n_k \sigma_k \bar{v}_k \tag{E.10}$$

Recall that $\sigma = \pi d^2$, where $d = (d_i + d_j)/2$. If we assume the molecular diameter is about 2.5 Å, then $\sigma \simeq 10^{-18}$ m^2. The relative collision speed at room temperature is about 300 m/sec. If $p = 1$ atm and $T = 300$ K, then $n = p/kT = 2.5 \times 10^{25}$ m^{-3} and $\nu \simeq 1.25 \times 10^{10}$ sec^{-1}. The inverse of ν is a characteristic relaxation or $1/e$ time, τ, for the first-order ordinary differential equation of Eq. (E.9). For this example, $\tau \simeq 10^{-10}$ sec or 0.1 nsec. Thus, in the absence of driving forces, the velocity distribution function rapidly relaxes to its equilibrium value under normal conditions. Of course, if the density is very small, then τ will increase accordingly and non-equilibrium effects become increasingly important.

In Appendix D we derived the multicomponent conservation equations. These equations are generally valid, regardless of the thermodynamic state and degree of non-equilibrium. However, they also require the simultaneous solution of the Boltzmann equation in order to determine the transport terms v_{k_i}, τ_{ij}, and q_i and in some domains direct solution of the conservation equations becomes intractable. Figure 12.1 illustrates the range of validity or applicability of various methods for describing flow. The most important distinction illustrated by this figure is that of continuum versus free molecular flow. In the limit of very small Knudsen number, the Euler equations apply. Known also as the inviscid conservation equations, in this limit collisional relaxation is so fast that the transport terms can be neglected. As the Knudsen number increases, the possibility of nonlinear effects increases, and the transport terms become important. If the Knudsen number isn't too large, however, then the departure of the distribution function from equilibrium, while small, is sufficient to lead to significant transport phenomena. Because the departure from equilibrium is small, expansion techniques can be used to solve Boltzmann's equation. Such methods were first shown by Chapman and Enskog and led to the Navier–Stokes equations in which there is a linear relationship between transport and the first derivatives of mass, momentum, or energy. For somewhat larger Knudsen numbers, higher-order relations are needed; the Burnett equations are an example. For Knudsen numbers beyond about unity, the analytical solution of

Boltzmann's equation becomes intractable and numerical methods are required. The most common techniques are discrete Monte Carlo simulation (DSMC) and molecular dynamics. However, numerous other approaches have been explored.

E.1 Collision Integrals

The general collision integral is

$$\Omega_{ij}^{(l,s)}(T) = \sqrt{\frac{kT}{2\pi\mu_{ij}}} \int_0^\infty e^{-\gamma^2} \gamma^{2s+3} Q^{(l)}(g) d\gamma \tag{E.11}$$

where

$$Q^{(l)}(g) = 2\pi \int_0^\infty (1 - \cos^l \chi) b db \tag{E.12}$$

and

$$\gamma^2 = \frac{1}{2}\frac{\mu_{ij} g^2}{kT} \tag{E.13}$$

It is usual to normalize Ω on the rigid sphere value, identifying the normalized value with a star:

Table E.1 Lennard–Jones 6,12 collision integrals (from Hirschfelder, Curtiss, and Bird [14]; also in Bird, Stewart, and Lightfoot [15])

T^*	$\Omega^{(2,2)*}$	$\Omega^{(1,1)*}$	T^*	$\Omega^{(2,2)*}$	$\Omega^{(1,1)*}$	T^*	$\Omega^{(2,2)*}$	$\Omega^{(1,1)*}$	T^*	$\Omega^{(2,2)*}$	$\Omega^{(1,1)*}$
0.3	2.785	2.662	1.3	1.399	1.273	2.6	1.081	0.9878	4.6	0.9422	0.8568
0.35	2.628	2.476	1.35	1.375	1.253	2.7	1.069	0.977	4.7	0.9382	0.853
0.4	2.492	2.318	1.4	1.353	1.233	2.8	1.058	0.9672	4.8	0.9343	0.8492
0.45	2.368	2.184	1.45	1.333	1.215	2.9	1.048	0.9576	4.9	0.9305	0.8456
0.5	2.257	2.066	1.5	1.314	1.198	3	1.039	0.949	5	0.9269	0.8422
0.55	2.156	1.966	1.55	1.296	1.182	3.1	1.03	0.9406	6	0.8963	0.8124
0.6	2.065	1.877	1.6	1.279	1.167	3.2	1.022	0.9328	7	0.8727	0.7896
0.65	1.982	1.798	1.65	1.264	1.153	3.3	1.014	0.9256	8	0.8538	0.7712
0.7	1.908	1.729	1.7	1.248	1.14	3.4	1.007	0.9186	9	0.8379	0.7556
0.75	1.841	1.667	1.75	1.234	1.128	3.5	0.999	0.912	10	0.8242	0.7424
0.8	1.78	1.612	1.8	1.221	1.116	3.6	0.9932	0.9058	20	0.7432	0.664
0.85	1.725	1.562	1.85	1.209	1.105	3.7	0.987	0.8998	30	0.7005	0.6232
0.9	1.675	1.517	1.9	1.197	1.094	3.8	0.9811	0.8942	40	0.6718	0.596
0.95	1.629	1.476	1.95	1.186	1.084	3.9	0.9755	0.8888	50	0.6504	0.5756
1	1.587	1.439	2	1.175	1.075	4	0.97	0.8836	60	0.6335	0.5596
1.05	1.549	1.406	2.1	1.156	1.057	4.1	0.9649	0.8788	70	0.6194	0.5464
1.1	1.514	1.375	2.2	1.138	1.041	4.2	0.96	0.874	80	0.6076	0.5352
1.15	1.482	1.346	2.3	1.122	1.026	4.3	0.9553	0.8694	90	0.5973	0.5256
1.2	1.452	1.32	2.4	1.107	1.012	4.4	0.9507	0.8652	100	0.5882	0.517
1.25	1.424	1.296	2.5	1.093	0.996	4.5	0.9464	0.861			

$$\Omega^{(l,s)*}(T^*) \equiv \frac{\Omega^{(l,s)}(T)}{\left[\Omega^{(l,s)}\right]_{\text{rigid sphere}}} = \frac{\Omega^{(l,s)}}{\frac{1}{2}(s+1)!\left[1 - \frac{1}{2}\frac{1+(-1)^l}{1+l}\right]\pi\sigma^2} \tag{E.14}$$

Here

$$T^* \equiv \frac{kt}{\epsilon} \tag{E.15}$$

ϵ and σ are potential energy parameters for the Lennard–Jones potential

$$\phi(r) = 2\epsilon\left[(\sigma/r)^{12} - (\sigma/r)^6\right] \tag{E.16}$$

Values for $\Omega^{(l,s)*}(T^*)$ are listed in Table E.1 for the Lennard–Jones 6,12 potential.

F Bibliography for Thermodynamics

The following annotated list of books related to equilibrium thermodynamics is provided as a guide and is not intended to be an exhaustive list of books on the subject.

Classical Thermodynamics

Cengel and Boles, *Thermodynamics: An Engineering Approach*, 8th edn [44]. The current gold standard in mechanical engineering undergraduate texts.

Callen, *Thermodynamics and an Introduction to Thermostatistics*, 2nd edn [3]. A complete development of a deductive axiomatic approach to thermodynamics is provided. Applications to chemical, electromagnetic, and solid systems are included. Some treatment of irreversible thermodynamics.

Fermi, *Thermodynamics* [68]. This paperback is notable primarily because it has an accurate derivation of the inequality $\delta Q/T \geq 0$.

Gibbs, *Elementary Principles in Statistical Mechanics* in *Collected Works* [4]. Classic book from which all modern thermodynamics flows.

Guggenheim, *Thermodynamics*, 5th edn [69]. This edition has been influenced by Callen. A good, short chapter on the connections between statistical and macroscopic thermodynamics. Stronger chemical emphasis than Callen.

Reynolds and Perkins, *Engineering Thermodynamics* [70]. One of the classics for introductory courses in engineering thermodynamics. Entropy is treated from a microscopic viewpoint.

Zemansky, *Heat and Thermodynamics*, 5th edn [19]. This is the classic introductory text to thermodynamics. A lucid and rigorous presentation of the inductive physical approach to thermodynamics, as formulated by Clausius and Kelvin, is presented.

Introductory Statistical Thermodynamics

Hill, *Introduction to Statistical Thermodynamics* [71]. This book was a standard for many years.

Incropera, *Introduction to Molecular Structure and Thermodynamics* [7]. Strong physical basis for statistical thermodynamics, including spectroscopy. Excellent presentation of quantum mechanics at a level understandable to mechanical engineers.

Knuth, *Introduction to Statistical Thermodynamics* [5]. This book provides a careful introduction to ensemble theory.

Reif, *Fundamentals of Statistical and Thermal Physics* [72]. Thermodynamics, statistical thermodynamics, and kinetic theory are provided a unified treatment. Some topics are treated at a more advanced level.

Laurendeau, *Statistical Thermodynamics: Fundamentals and Applications* [73]. A more recent, well-written text. Focuses more on classical statistics. Excellent chapters on spectroscopy.

More Advanced Statistical Mechanics and Applications

Hirschfelder, Curtiss, and Bird, *Molecular Theory of Gases and Liquids* [14]. This 1275-page book is a classic on intermolecular forces and applications to theoretical prediction of non-ideal gas behavior and transport properties. Commonly referred to as "HCB," "MTGL," or "The Great Green Bible." It has pretty much everything on the subject.

McQuarrie, *Statistical Mechanics* [31]. An excellent book with many modern developments at a higher level than we explore.

Quantum Mechanics

Pauling and Wilson, *Introduction to Quantum Mechanics with Applications to Chemistry* [23]. The standard chemistry-oriented text on quantum mechanics that is still easy to read.

Shankar, *Principles of Quantum Mechanics*, 2nd edn [74]. Commonly used in first-year graduate quantum mechanics courses in chemistry.

Cohen-Tannoudji, Diu, and Laloe, *Quantum Mechanics*, Vols 1&2 [75]. More detailed that Shankar, also used in graduate quantum mechanics courses in chemistry.

Messiah, *Quantum Mechanics* Vols 1&2 [76]. The bible of quantum mechanics. You will need to take a real quantum mechanics course to understand it.

Eisenbud, *The Conceptual Foundations of Quantum Mechanics* [77]. A good way to get a larger feel for quantum mechanics.

Gamow and Stannard, *The New World of Mr Tompkins: George Gamow's Classic Mr Tompkins in Paperback* [78]. Humorous wanderings with Mr Tompkins into the world of relativity and quantum mechanics.

Atoms and Molecules

The following books by Gerhard Herzberg are the most complete works on atomic and molecular structure and spectra:

Atomic Spectra and Atomic Structure [79].

Molecular Spectra and Molecular Structure I: Diatomic Molecules [28].

Molecular Spectra and Molecular Structure II: Infrared and Raman Spectra of Polyatomic Molecules [80].

Molecular Spectra and Molecular Structure III: Electronic Spectra and Electronic Structure of Polyatomic Molecules [29].

The Spectra and Structure of Simple Free Radicals [81].

More on atoms and molecules:

Radzig and Smirov, *Reference Data on Atoms, Molecules and Ions* [9]. Lots of useful data on atoms and molecules.

Liquids

Allen and Tildesley, *Computer Simulation of Liquids* [82].

Galiatsatos (Ed.), *Molecular Simulation Methods for Predicting Polymer Properties* [83].

Molecular Dynamics and Computational Chemistry

Cramer, *Essentials of Computational Chemistry*, 2nd edn [84].

Haile, *Molecular Dynamics Simulation* [85].

Frenkela and Smit, *Understanding Molecular Simulation – From Algorithms to Applications* [86].

Schlick, *Molecular Modeling and Simulation – An Interdisciplinary Guide* [87].

Kinetic Theory and Transport Phenomena

Jeans, *An Introduction to the Kinetic Theory of Gases* [88].

Kennard, *Kinetic Theory of Gases: With an Introduction to Statistical Mechanics* [89].

Vincenti and Kruger, *Introduction to Physical Gas Dynamics* [2].

Bird, Stewart, and Lightfoot, *Transport Phenomena*, 2nd edn [15]. A very useful book for solving problems involving transport of mass, momentum, or heat. A bible for chemical engineers.

Chapman and Cowling, *The Mathematical Theory of Non-Uniform Gases*, 3rd edn [46]. The origin of modern kinetic theory.

Useful Mathematics References

Abramowitz and Stegun (eds), *Handbook of Mathematical Functions with Formulas, Graphs, and Mathematical Tables* [27].

Online Resources

NIST–JANNAF Thermochemical Tables. The central source for thermochemical properties is the JANNAF tables. This is a compilation of properties by the National Institute of Standards and Technology. The data is presented in a standard format that is used by many computer programs (http://kinetics.nist.gov/janaf/).

NIST Chemistry Webbook. The webbook is an extensive compilation of chemical and physical property data (http://webbook.nist.gov/chemistry/).

NIST Chemical Kinetics Database. An extensive database of chemical reaction rates (http://kinetics.nist.gov/kinetics/index.jsp).

Argonne Active Thermochemical Tables. Active Thermochemical Tables (ATcT) provide thermochemical values (such as enthalpies of formation, Gibbs energies of formation, bond dissociation energies, reaction enthalpies, etc.) for stable, reactive, and transient chemical species (http://atct.anl.gov).

Some Useful Journals

AIAA journals (several)
AIChE Journal
Applied Optics
Applied Physics Letters
ASHRAE Transactions
ASME journals (several)
Chemical Physics Letters
Cryogenics
Heating, Piping and Air Conditioning
High Temperatures–High Pressures
Industrial and Engineering Chemistry
Metallurgical Transactions A
Journal of the American Ceramics Society
Journal of Applied Physics
Journal of Applied Polymer Science

Journal of Chemical Physics
Journal of Metals
Journal of Physical Chemistry
Journal of the Optical Society of America
Journal of Quantitative Spectroscopy and Radiative Transfer
Physics of Fluids
Physics Letters
Physical Review Letters
Physical Reviews
Reviews of Modern Physics

References

[1] R. E. Sonntag and G. J. V. Wylen, *Fundamentals of Statistical Thermodynamics*, Wiley, Chichester, 1966.

[2] W. G. Vincenti and J. C. H. Kruger, *Introduction to Physical Gas Dynamics*, Wiley, Chichester, 1965.

[3] H. B. Callen, *Thermodynamics and an Introduction to Thermostatistics*, 2nd edn, Wiley, Chichester, 1985.

[4] J. Gibbs, *Collected Works*, Yale University Press, New Haven, CT, 1948.

[5] E. Knuth, *Introduction to Statistical Thermodynamics*, McGraw-Hill, New York, 1966.

[6] C. L. Tien and J. H. Lienhard, *Statistical Thermodynamics*, McGraw-Hill, New York, 1971.

[7] F. P. Incropera, *Introduction to Molecular Structure and Thermodynamics*, Wiley, Chichester, 1974.

[8] R. Perry and D. Green, *Perry's Chemical Engineers' Handbook*, McGraw-Hill, New York, 1984.

[9] A. Radzig and B. Smirov, *Reference Data on Atoms, Molecules and Ions*, Springer-Verlag, New York, 1985.

[10] C. Kittel, *Solid State Physics*, 2nd edn, Wiley, Chichester, 1956.

[11] G. White and S. Collocott, Heat capacity of reference materials: Cu and W, *Journal of Physical Chemistry Reference Data* 13(4) (1984) 1251–1257.

[12] G. Bird, *Molecular Gas Dynamics and the Direct Simulation of Gas Flows*, 2nd edn, Oxford University Press, Oxford, 1976.

[13] J. Li, Z. Zhao, A. Kazakov, and F. L. Dryer, An updated comprehensive kinetic model of hydrogen combustion, *International Journal of Chemical Kinetics* 36(10) (2004) 566–575.

[14] J. Hirschfelder, C. Curtiss, and R. Bird, *Molecular Theory of Gases and Liquids*, Wiley, Chichester, 1964.

[15] R. B. Bird, W. E. Stewart, and E. N. Lightfoot, *Transport Phenomena*, 2nd edn, Wiley, New York, 2007.

[16] J. van der Waals, Over de continuiteit van den gas- en vloeistoftoestand (on the continuity of the gas and liquid state), PhD thesis, Leiden, The Netherlands (1873).

[17] R. Alberty, Use of Legendre transformations in chemical thermodynamics, *Pure and Applied Chemistry* 73(8) (2001) 1349–1380.

[18] B. Mahon, *The Man Who Changed Everything: The Life of James Clerk Maxwell*, Wiley, New York, 2007.

[19] M. Zemansky, *Heat and Thermodynamics*, 5th edn, McGraw-Hill, New York, 1960.

[20] E. W. Lemmon, M. L. Huber, and M. O. McLinden, NIST standard reference database 23: Reference fluid thermodynamic and transport properties–refprop, version 9.1, National Institute of Standards and Technology, https://www.nist.gov/srd/refprop (2013).

[21] H. Robbins, A remark on Stirling's formula, *American Mathematics Monthly* 62 (1955) 26–29.

[22] R. Courant and D. Hilbert, *Methods of Mathematical Physics*, Vol. 2, Wiley, New York, 2008.

[23] L. Pauling and E. Wilson, *Introduction to Quantum Mechanics with Applications to Chemistry*, McGraw-Hill, New York, 1935.

[24] R. I. Gamow, *Thirty Years that Shook Physics: The Story of Quantum Theory*, Dover, Mineola, NY, 1985.

[25] J. Burnet, *Early Greek Philosphy*, Kessinger Publishing, Whitefish, MT, 2003.

[26] J. Whiting and M. Kjelle, *John Dalton and the Atomic Theory (Uncharted, Unexplored, and Unexplained)*, Mitchell Lane Publishers, Newark, DE, 2004.

[27] M. Abramowitz and I. A. Stegun (Eds), *Handbook of Mathematical Functions with Formulas, Graphs, and Mathematical Tables*, Dover, New York, 1965.

[28] G. Herzberg, *Molecular Spectra and Molecular Structure I: Diatomic Molecules*, Prentice Hall, New York, 1939.

[29] G. Herzberg, *Molecular Spectra and Molecular Structure III: Electronic Spectra and Electronic Structure of Polyatomic Molecules*, Van Nostrand Reinhold, New York, 1966.

[30] J. Luque and D. Crosley, LIFBASE: Database and spectral simulation (version 1.5). Tech. Rep. Report MP 99-009, SRI International (1999).

[31] D. McQuarrie, *Statistical Mechanics*, Harper and Row, New York, 1976.

[32] M. W. Chase, Jr., NIST–JANAF thermochemical tables, *Journal of Physical Chemistry Reference Data Monograph* 9 (1998) 1–1951.

[33] B. McBride and S. Gordon, Computer program for calculation of complex chemical equilibrium compositions and applications II. User's manual and program description, Tech. Rep. NASA RP-1311-P2, NASA (1996).

[34] ANSYS CHEMKIN Pro 18.1, http://www.ansys.com/products/fluids/combustion-tools.

[35] O. Redlich and J. Kwong, On the thermodynamics of solutions; an equation of state; fugacities of gaseous solutions, *Chemical Reviews* 44 (1949) 233.

[36] D. Peng and D. Robinson, A new two-constant equation of state, *Industrial Engineering Chemical Engineering Fundamentals* 15 (1976) 59.

[37] M. Benedict, G. Webb, and L. Rubin, An empirical equation for thermodynamic properties of light hydrocarbons and their mixtures I. Methane, ethane, propane and *n*-butane, *Journal of Chemical Physics* 8 (1940) 334.

[38] M. Benedict, G. Webb, and L. Rubin, An empirical equation for thermodynamic properties of light hydrocarbons and their mixtures II. Mixtures of methane, ethane, propane, and *n*-butane, *Journal of Chemical Physics* 10 (1942) 747.

[39] H. Cooper and J. Goldfrank, constants and new correlations, *Hydrocarbon Processing* 46(12) (1967) 141–146.

[40] S. Plimpton, Fast parallel algorithms for short-range molecular dynamics, *Journal of Computational Physics* 117 (1995) 1–19.

[41] I. Todorov, W. Smith, K. Trachenko, and M. Dove, Dl_POLY_3: New dimensions in molecular dynamics simulations via massive parallelism, *Journal of Material Chemistry* 16 (2006) 1911–1918.

[42] L. Lagardère, L.-H. Jolly, F. Lipparini, F. Aviat, B. Stamm, Z. F. Jing et al., Tinker-hp: A massively parallel molecular dynamics package for multiscale simulations of large complex systems with advanced point dipole polarizable force fields, *Chemical Science* 9 (2018) 956–972.

[43] A. Van Itterbeek and O. Verbeke, Density of liquid nitrogen and argon as a function of pressure and temperature, *Physica* 26(11) (1960) 931–938.

[44] Y. A. Cengel and M. A. Boles, *Thermodynamics: An Engineering Approach*, 8th edn, McGraw-Hill, New York, 2014.

[45] E. Eucken, Ueber das wärmeleitvermogen, die spezifische wärme und die innere reibung der gase, *Physikalische Zeitschrift* (1913) 324–332.

[46] S. Chapman and T. G. Cowling, *The Mathematical Theory of Non-Uniform Gases*, 3rd edn, Cambridge University Press, Cambridge, 1970.

[47] F. A. Williams, *Combustion Theory*, 2nd edn, Addison-Wesley, New York, 1988.

[48] H. Petersen, The properties of helium: Density, specific heats, viscosity, and thermal conductivity at pressures from 1 to 100 bar and from room temperature to about 1800 K. Report 224, Danish Atomic Energy Commission (1970).

[49] J. Kestin, K. Knierim, E. A. Mason, B. Najafi, S. T. Ro, and M. Waldman, Equilibrium and transport properties of the noble gases and their mixtures at low density, *Journal of Physical and Chemical Reference Data* 13(1) (1984) 229.

[50] R. Hanson, R. Spearrin, and C. Goldenstein, *Spectroscopy and Optical Diagnostics for Gases*, Springer, Berlin, 2016.

[51] A. Eckbreth, *Laser Diagnostics for Combustion Temperature and Species*, 2nd edn, Gordon & Breach, Philadelphia, PA, 1996.

[52] A. Einstein, The quantum theory of radiation, *Physikalische Zeitschrift* 18 (1917) 121.

[53] J. E. Sansonettia and W. C. Martin, Handbook of basic atomic spectroscopic data, *Journal of Physical and Chemical Reference Data* 34 (2005) 1559–2259.

[54] A. Mitchell and M. Zemansky, *Resonance Radiation and Excited Atom*, Cambridge University Press, Cambridge, 1961.

[55] J. Humlicek, An efficient method for evaluation of the complex probability function: The Voigt function and its derivatives, *Journal of Quantitative Spectroscopy and Radiative Transfer* 21 (1979) 309.

[56] J. I. Steinfeld and J. S. Francisco, *Chemical Kinetics and Dynamics*, Prentice Hall, Englewood Cliffs, NJ, 1998.

[57] G. Marin and G. Yablonsky, *Kinetics of Chemical Reactions*, Wiley, New York, 2011.

[58] P. Houston, *Chemical Kinetics and Reaction Dynamics*, McGraw Hill, New York, 2001.

[59] S. Arrhenius, Über die dissociationswärme und den einfluß der temperatur auf den dissociationsgrad der elektrolyte, *Zeitschrift für Physikalische Chemie* 4 (1889) 98–116.

[60] R. G. Gilbert and S. C. Smith, *Theory of Unimolecular and Recombination Reactions*, Blackwell Scientific, Oxford, 1990.

[61] R. Sivaramakrishnan, J. V. Michael, and S. J. Klippenstein, Direct observation of roaming radicals in the thermal decomposition of acetaldehyde, *Journal of Physical Chemistry A* 114(2) (2010) 755–764.

[62] L. B. Harding, Y. Georgievskii, and S. J. Klippenstein, Roaming radical kinetics in the decomposition of acetaldehyde, *Journal of Physical Chemistry A* 114(2) (2010) 765–777.

[63] G. P. Smith, D. M. Golden, M. Frenklach, N. W. Moriarty, B. Eiteneer, M. Goldenberg et al. GRI-Mech version 3.0, http://www.me.berkeley.edu/gri_mech/.

[64] D. G. Goodwin, H. K. Moffat, and R. L. Speth, Cantera: An object-oriented software toolkit for chemical kinetics, thermodynamics, and transport processes, http://www.cantera.org, version 2.3.0 (2017).

[65] H. Hippler, J. Troe, and H. J. Wendelken, Collisional deactivation of vibrationally highly excited polyatomic molecules. II. Direct observations for excited toluene, *The Journal of Chemical Physics* 78(11) (1983) 6709–6717.

[66] F. M. Mourits and F. H. A. Rummens, A critical evaluation of the Lennard–Jones and Stockmayer potential parameters and of some correlation methods, *Canadian Journal of Chemistry* 55 (1977) 3007.

[67] G. B. Ellison, Private communication (2017).

[68] E. Fermi, *Thermodynamics*, Dover, New York, 1936.

[69] E. Guggenheim, *Thermodynamics*, 5th edn, Wiley, Chichester, 1967.

[70] W. Reynolds and H. Perkins, *Engineering Thermodynamics*, McGraw-Hill, New York, 1970.

[71] T. Hill, *Introduction to Statistical Thermodynamics*, Addison-Wesley, New York, 1960.

[72] F. Reif, *Fundamentals of Statistical and Thermal Physics*, Waveland Press, Long Grove, IL, 2008.

[73] N. M. Laurendeau, *Statistical Thermodynamics: Fundamentals and Applications*, Cambridge University Press, Cambridge, 2005.

[74] P. Shankar, *Principles of Quantum Mechanics*, 2nd edn, Plenum Press, New York, 2010.

[75] C. Cohen-Tannoudji, B. Diu, and F. Laloe, *Quantum Mechanics*, Vols 1&2, Wiley, New York, 2006.

[76] A. Messiah, *Quantum Mechanics*, Vols 1&2, Interscience, Geneva, 1961.

[77] L. Eisenbud, *The Conceptual Foundations of Quantum Mechanics*, AMS Chelsea Publishing, Providence, RI, 2007.

[78] G. Gamow and R. Stannard, *The New World of Mr Tompkins: George Gamow's Classic Mr Tompkins in Paperback*, Cambridge University Press, Cambridge, 1965.

[79] G. Herzberg, *Atomic Spectra and Atomic Structure*, Dover, New York, 1944.

[80] G. Herzberg, *Molecular Spectra and Molecular Structure II: Infrared and Raman Spectra of Polyatomic Molecules*, Van Nostrand Reinhold, New York, 1945.

[81] G. Herzberg, *The Spectra and Structure of Simple Free Radicals*, Cornell University Press, Ithaca, NY, 1971.

[82] M. Allen and D. Tildesley, *Computer Simulation of Liquids*, Oxford Science, Oxford, 1987.

[83] V. Galiatsatos (Ed.), *Molecular Simulation Methods for Predicting Polymer Properties*, Wiley-Interscience, Hoboken, NJ, 2005.

[84] C. Cramer, *Essentials of Computational Chemistry*, 2nd edn, Wiley, New York, 2006.

[85] J. Haile, *Molecular Dynamics Simulation*, Wiley, Chichester, 1997.

[86] D. Frenkela and B. Smit, *Understanding Molecular Simulation – From Algorithms to Applications*, Academic Press, New York, 2002.

[87] T. Schlick, *Molecular Modeling and Simulation – An Interdisciplinary Guide*, Springer-Verlag, New York, 2002.

[88] J. Jeans, *An Introduction to the Kinetic Theory of Gases*, Cambridge University Press, Cambridge, 2009.

[89] E. H. Kennard, *Kinetic Theory of Gases: With an Introduction to Statistical Mechanics*, McGraw-Hill, New York, 1938.

Index